T0292593

Smart Innovation, Systems and Technologies

Volume 46

Series editors

Robert J. Howlett, KES International, Shoreham-by-Sea, UK
e-mail: rjhowlett@kesinternational.org

Lakhmi C. Jain, University of Canberra, Canberra, Australia, and
Bournemouth University, UK
e-mail: Lakhmi.Jain@canberra.edu.au

About this Series

The Smart Innovation, Systems and Technologies book series encompasses the topics of knowledge, intelligence, innovation and sustainability. The aim of the series is to make available a platform for the publication of books on all aspects of single and multi-disciplinary research on these themes in order to make the latest results available in a readily-accessible form. Volumes on interdisciplinary research combining two or more of these areas is particularly sought.

The series covers systems and paradigms that employ knowledge and intelligence in a broad sense. Its scope is systems having embedded knowledge and intelligence, which may be applied to the solution of world problems in industry, the environment and the community. It also focusses on the knowledge-transfer methodologies and innovation strategies employed to make this happen effectively. The combination of intelligent systems tools and a broad range of applications introduces a need for a synergy of disciplines from science, technology, business and the humanities. The series will include conference proceedings, edited collections, monographs, handbooks, reference books, and other relevant types of book in areas of science and technology where smart systems and technologies can offer innovative solutions.

High quality content is an essential feature for all book proposals accepted for the series. It is expected that editors of all accepted volumes will ensure that contributions are subjected to an appropriate level of reviewing process and adhere to KES quality principles.

More information about this series at http://www.springer.com/series/8767

Ioannis Hatzilygeroudis · Vasile Palade
Jim Prentzas
Editors

Combinations of Intelligent Methods and Applications

Proceedings of the 4th International Workshop, CIMA 2014, Limassol, Cyprus, November 2014 (at ICTAI 2014)

 Springer

Editors
Ioannis Hatzilygeroudis
Department of Computer Engineering
  and Informatics, School of Engineering
University of Patras
Patras
Greece

Jim Prentzas
Laboratory of Informatics, School
  of Education Sciences, Department
  of Education Sciences in Early Childhood
Democritus University of Thrace
Alexandroupolis
Greece

Vasile Palade
Faculty of Engineering and Computer
  Science
Coventry University
Coventry
UK

ISSN 2190-3018   ISSN 2190-3026 (electronic)
Smart Innovation, Systems and Technologies
ISBN 978-3-319-26858-3   ISBN 978-3-319-26860-6 (eBook)
DOI 10.1007/978-3-319-26860-6

Library of Congress Control Number: 2015958315

Printed on acid-free paper

This Springer imprint is published by SpringerNature
The registered company is Springer International Publishing AG Switzerland

Preface

The combination of different intelligent methods is an active research area in Artificial Intelligence (AI). The aim is to create integrated or hybrid methods that benefit from each of their components. It is generally believed that complex problems can be easier solved with such integrated or hybrid methods.

Some of the existing efforts combine what are called soft computing methods (fuzzy logic, neural networks and genetic algorithms) either among themselves or with more traditional AI methods such as logic and rules. Another stream of efforts integrates case-based reasoning or machine learning with soft computing or traditional AI methods. Yet another integrates agent-based approaches with logic and also non-symbolic approaches. Some of the combinations have been quite important and more extensively used, like neuro-symbolic methods, neuro-fuzzy methods and methods combining rule-based and case-based reasoning. However, there are other combinations that are still under investigation, such as those related to the Semantic Web. In some cases, combinations are based on first principles, whereas in other cases they are created in the context of specific applications.

The 4th Workshop on "Combinations of Intelligent Methods and Applications" (CIMA 2014) was intended to become a forum for exchanging experience and ideas among researchers and practitioners dealing with combining intelligent methods either based on first principles or in the context of specific applications.

Important issues of the Workshop were (but not limited to) the following:

- Case-Based Reasoning Integrations
- Genetic Algorithms Integrations
- Combinations for the Semantic Web
- Combinations and Web Intelligence
- Combinations and Web Mining
- Fuzzy-Evolutionary Systems
- Hybrid deterministic and stochastic optimisation methods
- Hybrid Knowledge Representation Approaches/Systems
- Hybrid and Distributed Ontologies
- Information Fusion Techniques for Hybrid Intelligent Systems

- Integrations of Neural Networks
- Intelligent Agents Integrations
- Machine Learning Combinations
- Neuro-Fuzzy Approaches/Systems
- Applications of Combinations of Intelligent Methods to

  - Biology and Bioinformatics
  - Education and Distance Learning
  - Medicine and Health Care

CIMA 2014 was held in conjunction with the 26th IEEE International Conference on Tools with Artificial Intelligence (ICTAI 2014).

This volume includes extended and revised versions of the papers presented in CIMA 2014.

We would like to express our appreciation to all authors who submitted papers as well as to the members of the CIMA-14 program committee for their excellent work. We would also like to thank the ICTAI 2014 PC Chair for accepting to host CIMA 2014.

We hope that this proceedings will be of benefit to both researchers and developers. Given the success of the first four Workshops on combinations of intelligent methods, we intend to continue our effort in the coming years.

Ioannis Hatzilygeroudis
Vasile Palade
Jim Prentzas

Organization Committee

Chairs-Organizers

Ioannis Hatzilygeroudis, University of Patras, Greece
Vasile Palade, University of Oxford, UK
Jim Prentzas, Democritus University of Thrace

Program Committee

Salha Alzahrani, Taif University, Saudi Arabia
Soumya Banerjee, Birla Institute of Technology, India
Nick Bassiliades, Aristotle University of Thessaloniki, Greece
Gloria Cerasela Crisan, University of Bacau, Romania
Kit Yan Chan, Curtin University of Technology, Australia
Wei Fang, Jiangnan University, China
Andreas Holzinger, Medical University of Graz, Austria
Constantinos Koutsojannis, TEI of Patras, Greece
Rahat Iqbal, Coventry University, UK
Victoria Lopes-Morales, University of Granada
George Magoulas, Birkbeck College, Univ. of London, UK
Ashih Mani, Dayalbagh Educational Institute, India
Toni Moreno, University Rovira i Virgili, Spain
Ciprian-Daniel Neagu, University of Bradford, UK
Camelia Pintea, Technical University of Cluj-Napoca, Romania
Roozbeh Razavi-Far, Universite Libre de Bruxelles, Belgium
David Sanchez, University Rovira i Virgili, Spain
Kyriakos Sgarbas, Unversity of Patras, Greece
Jun Sun, Jiangnan University, China
George Tsihrintzis, University of Piraeus, Greece

Juan Velasquez, University of Chile, Chile
Douglas Vieira, Enacom-Handcrafted Technologies, Brazil

Contact Chair

Ioannis Hatzilygeroudis, Department of Computer Engineering and Informatics, University of Patras, Greece, e-mail: ihatz@ceid.upatras.gr

Contents

Evolutionary Landscape and Management of Population Diversity

Maumita Bhattacharya

Abstract   The search ability of an Evolutionary Algorithm (EA) depends on the variation among the individuals in the population [1–3]. Maintaining an optimal level of diversity in the EA population is imperative to ensure that progress of the EA search is unhindered by premature convergence to suboptimal solutions. Clearer understanding of the concept of population diversity, in the context of evolutionary search and premature convergence in particular, is the key to designing efficient EAs. To this end, this paper first presents a brief analysis of the EA population diversity issues. Next we present an investigation on a counter-niching EA technique [2] that introduces and maintains constructive diversity in the population. The proposed approach uses informed genetic operations to reach promising, but unexplored or under-explored areas of the search space, while discouraging premature local convergence. Simulation runs on a suite of standard benchmark test functions with Genetic Algorithm (GA) implementation shows promising results.

1   Introduction

Implementation of evolutionary algorithm (EA) requires preserving a population that maintains a degree of population diversity, while converging to a solution [3–10] in order to avoid premature convergence to sub-optimal solutions. It is difficult to precisely characterize the possible extent of premature convergence as it may occur in EA due to various reasons. The primary causes are algorithmic features like *high selection pressure* and *very high gene flow* among population members. Selection pressure pushes the evolutionary process to focus more and more on the already discovered better performing regions or "peaks" in the search space and as a result population diversity declines, gradually reaching a homogeneous state. On the other hand, unrestricted recombination results in high *gene flow* which spreads

M. Bhattacharya (✉)
School of Computing & Mathematics, Charles Sturt University,
Albury, NSW 2640, Australia
e-mail: mbhattacharya@csu.edu.au

© Springer International Publishing Switzerland 2016
I. Hatzilygeroudis et al. (eds.), *Combinations of Intelligent Methods
and Applications*, Smart Innovation, Systems and Technologies 46,
DOI 10.1007/978-3-319-26860-6_1

genetic material across the population, pushing it to a homogeneous state. Variation introduced through mutation is unlikely to be adequate to escape local optimum or optima [11]. While *premature convergence* [11] may be defined as the phenomenon of convergence to sub-optimal solutions, *gene-convergence* means loss of diversity in the process of evolution. Though, the convergence to a local or to the global optimum cannot necessarily be concluded from gene convergence, maintaining a certain degree of diversity is widely believed to help avoid entrapment in non-optimal solutions [1, 2].

In this paper we present an analysis on population diversity in the context of efficiency of evolutionary search. We then present an investigation on a counter niching-based EA that aims at combating gene-convergence (and premature convergence in turn) by employing intelligent introduction of constructive diversity [2].

The rest of the paper is organized as follows: Sect. 2 presents an analysis of diversity issues and the EA search process; Sect. 3 introduces the problem space for our proposed algorithm. Sections 4–6 present the proposed algorithm, simulation details and discussions on the results respectively. Finally, Sect. 7 presents some concluding remarks.

2 Population Diversity and Evolutionary Search

The EA search process depends on the variation among the individuals or candidate solutions in the population. In case of genetic algorithm and similar EAs, the variation is introduced by the *recombination* operator combining existing solutions, and the *mutation* operator introducing noise by applying random variation to the individual's genome. However, as the algorithm progresses, loss of diversity or loss of genetic variation in the population results in low exploration, pushing the algorithm to converge prematurely to a local optimum or non-optimal solution. Exploration in this context means searching new regions in the solution space; whereas, exploitation means performing searchs in the neighbourhoods which have been already visited. Success of the EA search process requires an optimal balance between exploitation and exploration.

In the context of EA, diversity may be described as the variation in the genetic material among individuals or candidate solutions in the EA population. This in turn may also mean variation in the fitness value of the individuals in the population. Two major roles played by population diversity in EA are as follows:

Firstly, diversity promotes exploration of the solution space to locate a single good solution by delaying convergence.

Secondly, diversity helps to locate multiple optima when more than one solution is present [3, 9, 10].

Besides the role of diversity regarding premature convergence in static optimization problems, diversity also seems to be beneficial in non-stationary environments. If the genetic material in the population is too similar, i.e., has converged towards single points in the search space, all future individuals will be trapped at that single point even though the optimal solution has moved on to another location in the fitness landscape. However, if the population is diverse, the mechanism of recombination will continue to generate new candidate solutions making it possible for the EA to discover new optima.

The following sub-section presents an analysis of the impact of population diversity on premature convergence, based on the concepts presented in [7].

2.1 Effect of Population Diversity on Premature Convergence

Let $\vec{X} = \left(X_{1,\ldots,X_N}\right) \in S^N$ be a population of individuals Y in the solution space S^N, where the population size is N; let $\vec{X}(0)$ be the initial population; $\mathbf{H}$ is a schema, i.e., a hyperplane of the solution space S. $\mathbf{H}$ may be represented by its defining components (defining *alleles*) and their corresponding values as $\mathbf{H}(a_{i1}, \ldots, a_{ik})$, where K ($1 \leq K \leq chromosome\ length$). Leung et al. in [7] have proposed the following measures related to population diversity in canonical genetic algorithm.

Degree of population diversity, $\delta\left(\vec{X}\right)$: Defined as the number of distinct components in the vector $\sum_{i=1}^{N} X_i$; and *Degree of population maturity*, $\mu\left(\vec{X}\right)$: Described as $\mu\left(\vec{X}\right) = l - \delta\left(\vec{X}\right)$ or the number of lost alleles.

With probability of mutation, $p(m) = 0$ and $\vec{X}(0) = \vec{X}_0$, according to Leung et al. [7] the following postulates hold true: For each solution, $Y \in \mathbf{H}\left(a_{i1}, \ldots, a_{i\mu(\vec{X}_0)};\ \vec{X}_0\right)$, there exists a $n \geq 0$ such that $Probability\left\{Y \in \vec{X}(n) / \vec{X}(0) = \vec{X}_0\right\} > 0$. Conversely, for each solution, $Y \notin \mathbf{H}\left(a_{i1}, \ldots, a_{i\mu(\vec{X}_0)};\ \vec{X}_0\right)$, and every $n \geq 0$ such that $Probability\left\{Y \in \vec{X}(n) / \vec{X}(0) = \vec{X}_0\right\} = 0$.

It is obvious from the above postulates that the search ability of a canonical genetic algorithm is confined to the minimum schema with $2^{\delta(\vec{X})}$ different individuals. Hence, the greater the degree of population diversity, $\delta\left(\vec{X}\right)$, the greater is the search ability of the genetic algorithm. Conversely, a small degree of population diversity will mean limited search ability, reducing to zero search ability with $\delta\left(\vec{X}\right) = 0$.

2.2 Enhanced EAs to Combat Diversity Issues

No mechanism in a standard EA guarantees that the population will remain diverse throughout the run [11, 12]. Although there is a wide coverage of the fitness landscape at initialization due to the random initialization of individuals' genomes, selection quickly eliminates the least fit solutions, which implies that the population will converge towards similar points or even single points in the search space. Since the

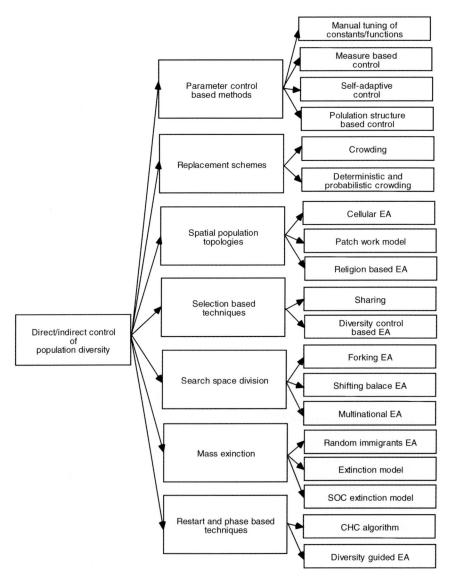

Fig. 1 Direct or indirect control of population diversity in EA

standard EA has limitations to maintain population diversity, several models have been proposed by the EA community which either maintain or reintroduce diversity in the EA population [2, 8, 13–19]. The key researches can be broadly categorized as follows [9]:

1. Complex population structures to control gene flow, e.g., the diffusion model, the island model, the multinational EA and the religion model.
2. Specialized operators to control and assist the selection procedure, e.g., crowding, deterministic crowding, and sharing are believed to maintain diversity in the population.
3. Reintroduction of genetic material, e.g., random immigrants and mass extinction models are aimed at reintroduction of diversity in the population.
4. Dynamic Parameter Encoding (DPE), which dynamically resizes the available range of each parameter by expanding or reducing the search window.
5. Diversity guided or controlled genetic algorithms that use a diversity measure to assess and control the survival probability of individuals and the process of exploration and exploitation.

Figure 1 summarizes the major methods proposed to directly or indirectly control EA population diversity.

The Counter-Niching based EA framework presented in this paper, employs a synergistic hybrid mechanism that combines the benefits of *specialized operator* and *reintroduction of diversity*.

3 Understanding the Problem Space

Before we present our proposed approach, which aims at achieving constructive diversity, it is important to understand the problem space we are dealing with. For optimization problems the main challenge is often posed by the topology of the fitness landscape, in particular its ruggedness in terms of local optima. The target optimization problems for our approach are primarily multimodal. Genetic diversity of the population is particularly important in case of multimodal fitness landscape. Evolutionary algorithms are required to avoid and escape local optima or basins of attraction to reach the optimum in a multimodal fitness landscape.

Over the years, several new and enhanced EAs have been suggested to improve performance [2, 8, 13–18, 20–24]. The objectives of much of this research are twofold; *firstly*, to avoid stagnation in local optimum in order to find the global optimum; *secondly*, to locate multiple good solutions if the application requires so.

In the second case, i.e., to locate multiple good solutions, alternative and different solutions may have to be considered before accepting one final solution as the optimum. An algorithm that can keep track of multiple optima simultaneously should be able to find multiple optima in the same run by spreading out the search.

On the other hand, maintaining genetic diversity in the population can be particularly beneficial in the first case; the problem of entrapment in local optima. Mutation is not sufficient to escape local optima as selection traditionally favours the better fit solutions entrapped in local optima. Genetic diversity is crucial as a diverse population allows the recombination operators to find different and newer solutions.

Remarks: The issue is—*how much genetic diversity in the population is optimum*?

Unfortunately, the answer to the above question is not a straightforward one because of the complex interplay among the variation and the selection operators as well as the characteristics of the problem itself. Recombination in a fully converged population cannot produce solutions that are different from the parents; let alone better than the parents. Interestingly, Ishibuchi et al. [25] used a NSGA-II implementation to demonstrate that similar parents actually improved diversity without adversely influencing convergence. A very high diversity on the other hand, actually deteriorates performance of the recombination operator. Offspring generated combining two parents approaching two different peaks is likely to be placed somewhere between the two peaks; hindering the search process from reaching either of the peaks. This makes the recombination operator less efficient for fine-tuning the solutions to converge at the end of the run. Hence, the optimal level of diversity is somewhere between fully converged and highly diverse. Various diversity measures (such as Euclidean distance among candidate solutions, fitness distance and so on) may be used to analyze algorithms to evaluate their diversity maintaining capabilities.

In the following sections we investigate the functioning and performance of our proposed Counter Niching-based Evolutionary Algorithm [2].

4 Counter Niching EA: The Operational Framework

To attain the objective of introducing constructive diversity in the population, the proposed technique first extracts information about the population landscape before deciding on introduction of diversity through informed mutation. The aim is to identify locally converging regions or *donor* communities in the landscape whose redundant less fit members (or individuals) could be replaced by more promising members sampled in un-explored or under-explored sections of the decision space. The existence of such communities is purely based on the position and spread of individuals in the decision space at a given point in time. Once such regions are identified, random sampling is done on yet to be explored sections of the landscape. Best representatives found during such sampling, now replace the worst members of the identified *donor* regions. Best representatives are the ones that are fitness wise the fittest and spatially the farthest. Here, average Euclidean distance from representatives of all already considered regions (stored in a "memory" array) is the measure for spatial distance. Regular mutation and recombination takes place in the population as a whole. The basic framework is as depicted in Fig. 2.

The task described in Fig. 2 is carried out by the following three procedures:

1. **Procedure COUNTER NICHING EA**: This is the main algorithm that calls the procedures GRID_NICHING and INFORMED_OP. Basically, COUNTER_ NICHING_EA has a very similar construct to a canonical genetic algorithm (see Fig. 3) except that the genetic operations (recombination and mutation) are performed via procedures GRID_NICHING and INFORMED_OP. Procedure

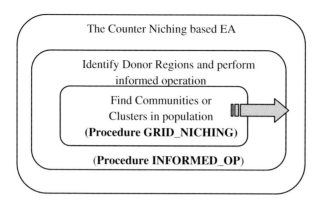

Fig. 2 The COUNTER NICHING based EA framework

Algorithm 1: Procedure COUNTER NICHING EA

1: **begin**
2: $t \leftarrow 0$
3: Initialize population $P(t)$
4: Evaluate population $P(t)$
5: **while** (not<termination condition>)
6: **begin**
7:     $t \leftarrow t+1$
    (* Perform pseudo-niching of the population*)
8:    Call Procedure GRID_NICHING
    (* Perform informed genetic operations *)
9:    Call Procedure INFORMED_OP
10:   Create new population using an elitist selection
     mechanism
11: Evaluate $P(t)$
14: **end while**
15: **end**

Fig. 3 The COUNTER NICHING based EA framework

GRID_NICHING is used to identify the formation of clusters or locally geno-typically converging regions in the solution space. Procedure INFORMED_OP, on the other hand, uses this clustering information to identify tendency towards fitness convergence, as this can be an early indication of premature convergence of the search process and hence, introduces diversity if necessary by a pseudo-mutation operator.

2. **Procedure GRID NICHING**: This function is called within COUNTER_ NICHING_EA and is used to identify local genotypic convergence. Here, we have used the term niching simply to connote identification of environments of individuals in the population, based on their genotypical information. In other words, we try to identify roughly the individual clusters in the decision space

based on their genotypic proximity. It may be noted that accuracy of the cluster boundaries is not of importance here. Instead, rough identification of cluster formation with reasonable amount of resources (runtime and memory space) is the prime objective.

Thus the procedure GRID-NICHING, returns information about community or cluster formation in the population, for the current generation.

3. **Procedure INFORMED OP**: The procedure INFORMED_OP is second in order to be called by COUNTER_NICHING_EA. This function is used for performing the genetic operations (recombination and mutation) along with an informed mutation in appropriate cases. The INFORMED_OP algorithm searches for locally converging communities with too many members of similar fitness. To achieve this, the clusters or regions in the list of "identified regions with high density" returned by GRID_NICHING are analyzed for potential fitness convergence. Redundant members of the high density clusters or regions with low fitness standard deviation (victim regions) are picked for replacement by promising members from relatively un-explored or under-explored sections (virgin zones) of the solution space. The idea is to explore greater parts of the solution space at the expense of these so-called redundant or extra members. We call this process informed mutation. The potential replacements are generated by random sampling of the solution space. A potential replacement thus generated is picked as actual replacement if it has fitness higher than the average fitness of the victim region and if it is furthest from all cluster centers compared to other candidates of similar fitness. However, *informed mutation* as explained above, thus operates on selected regions or communities only. Regular mutation and recombination is performed as usual on the entire population.

Figure 3 presents the procedure COUNTER_NICHING_EA. For details on the procedures GRID NICHING and INFORMED OP, we refer to our previous work in [2].

5   Simulations

5.1   Test Functions

Following the standard practice in the evolutionary computation research community, we have tested the proposed algorithm on a set of commonly used benchmark test functions to validate its efficacy.

The benchmark test function set used in the simulation runs consists of minimization of seven analytical functions given in Table 1: Ackley's Path Function $(f_{ack}(x))$, Griewank's Function $f_{gri}(x)$, Rastrigin's Function $f_{rtg}(x)$, Generalized Rosenbrock's function $f_{ros}(x)$, Axis parallel Hyper-Ellipsoidal Function or Weighted Sphere Model $f_{elp}(x)$, Schwefel Function 1.2 $f_{sch-1.2}(x)$ and a rotated Rastrigin Function $f_{rrtg}(x)$.

Table 1 Description of test functions

Function	Type	Global minimum
$f_{ack}(x) = 20 + e - 20\exp\left(-0.2\sqrt{\frac{1}{n}\sum_{i=1}^{n}x_i^2}\right)$ $-\exp\left(\frac{1}{n}\sum_{i=1}^{n}\cos\left(2\pi \cdot x_i\right)\right)$ $\quad -30 \leq x_i \leq 30$ where	Multimodal	$f_{ack}(x = 0) = 0$
$f_{gri}(x) = \frac{1}{4000}\sum_{i=1}^{n}(x_i - 100)^2 -$ $\prod_{i=1}^{n}\cos\left(\frac{x_i - 100}{\sqrt{i}}\right) + 1 \quad -600 \leq x_i \leq 600$ where	Multimodal, medium epistasis	$f_{gri}(x = 0) = 0$
$f_{rtg}(x) = \sum_{i=1}^{n}\left(x_i^2 - 10\cos\left(2\pi x_i\right) + 10\right)$ where $-5.12 \leq x_i \leq 5.12$	Multimodal, no epistasis	$f_{rtg}(x = 0) = 0$
$f_{ros}(x) = \sum_{i=1}^{n-1}\left(100\left(x_{i+1} - x_i^2\right)^2 + (x_i - 1)^2\right)$ where $-100 \leq x_i \leq 100$	Unimodal, high epistasis	$f_{ros}(x = 1) = 0$
$f_{elp}(x) = \sum_{i=1}^{M}ix_i^2 \quad -5.12 \leq x_i \leq 5.12$ where	Unimodal	$f_{elp}(x = 0) = 0$
$f_{sch-1.2}(x) = \sum_{i=1}^{M}\left(\sum_{k=1}^{i}x_k\right)^2$ where $-564 \leq x_i \leq 64$	Unimodal, high epistasis	$f_{sch-1.2}(x = 0) = 0$
$f_{rrtg}(x) = 10M + \sum_{i=1}^{M}\left(y_i^2 - 10\cos\left(2\pi y_i\right)\right)$ where $y = Ax$ with $A_{i,i} = 4/5,$ $A_{i,i+1} = 3/5_{(i\ odd)},$ $A_{i,i-1} = -3/5_{(i\ even)},$ $(A_{i,k} = 0_{(the\ rest)}$	Multimodal	$f_{rrtg}(x = 0) = 0$

Schwefel's function 1.2 and Rosenbrock's function are unimodal functions, but they have a strong epistasis among their variables. Griewank's function has very small but numerous minima around the global minimum, although it has a unimodal shape on a large scale. Rastrigin's function also has many local minima. However, it has no epistasis among its variables.

5.2 Algorithms Considered for Comparison

The algorithms used in the comparison are as follows:

1. The "standard EA" (SEA)
2. The self organized criticality EA (SOCEA)

3. The cellular EA (CEA), and
4. The diversity guided EA (DGEA)

The SEA uses Gaussian mutation with zero mean and variance $\sigma^2 = 1 + \sqrt{t+1}$. The SOCEA is a standard EA with non-fixed and non-decreasing variance $\sigma^2 = POW$ (10), where POW (α) is the power-law distribution. The purpose of the SOC-mutation operator is to introduce many small, some mid-sized, and a few large mutations. The effect of this simple extension is quite outstanding considering the effort to implement it in terms of lines of codes. The reader is referred to [9] for additional information on the SOCEA. Further, the CEA uses a 20×20 grid with wrapped edges. The grid size corresponds to the 400 individuals used in the other algorithms. The CEA uses Gaussian mutation with variance $\sigma^2 = POW(10)$, which allows comparison between the SOCEA and this version of the CEA. Mating is performed between the individual at a cell and a random neighbour from the four-neighbourhood. The offspring replaces the center individual if it has a better fitness than the center individual. Finally, the DGEA uses the Gaussian mutation operator with variance $\sigma^2 = POW$ (1). The diversity boundaries were set to $d_{low} = 5 \times 10^{-6}$ and $d_{high} = 0.25$, which proved to be good settings in preliminary experiments.

5.3 Experiment Set-Up

Simulations were carried out to apply the proposed COUNTER NICHING based EA with real-valued encoding with parameters N (population size) $= 300$, p_m (mutation probability) $= 0.01$ and p_r (recombination probability) $= 0.9$. In case of the algorithms used for comparison as mentioned in Sect. 5.2, namely, (i) SEA (Standard EA), (ii) SOCEA (Self-organized criticality EA), (iii) CEA (The Cellular EA), and (iv) DGEA (Diversity guided EA), experiments were performed using real-valued encoding, a population size of 400 individuals, and binary tournament selection. Probability of mutating an entire genome was $p_m = 0.75$ and probability for crossover was $p_r = 0.9$. As mentioned in Sect. 5.2, CEA uses a 20×20 grid with wrapped edges, where the grid size corresponds to the population size of 400 individuals as used in the other algorithms. The compared algorithms all use variants of the standard Gaussian mutation operator. The algorithm uses an arithmetic crossover with one weight for each variable. All weights except one are randomly assigned to either 0 or 1. The remaining weight is set to a random value between 0 and 1.

All the test functions were considered in 20, 50 and 100 dimensions. Reported results were averaged over 30 independent runs, maximum number of generations in each run being only 500, as against 1000 generations in used [9] for the same set of test cases for the 20 dimensional scenarios. The comparison algorithms use 50 times the dimensionality of the test problems as the terminating generation number in general, while the COUNTER NICHING EA uses 500, 1000 and 2000 generations for the 20, 50 and 100 dimensional problem variants respectively.

All the simulation processes were executed using a Pentium®4, 2.4 GHz CPU processor.

6   Results and Discussions

This section presents the empirical results obtained by the COUNTER NICHING EA algorithm when tackling the seven test problems mentioned in Sect. 5.1 with dimensions 20, 50 and 100.

6.1   General Performance of COUNTER NICHING EA

Table 2 presents the error values, $(f(x) - f(x)^*)$ where, $f(x)^*$ is the optimum. Each column corresponds to a test function. The error values have been presented for the three dimensions of the problems considered, namely 20, 50 and 100.

As each test problem was simulated over 30 independent runs, we have recorded results from each run and sorted the results in ascending order. Table 2 presents results from the representative runs: 1st (Best), 7th, 15th (Median), 22nd and 30th (Worst), Mean and Standard Deviation (Std). The main performance measures used are the following:

"A" **Performance**: Mean performance or average of the best-fitness function found at the end of each run. (Represented as 'Mean' in Table 2).

"SD" **Performance**: Standard deviation performance. (Represented as 'Std.' in Table 2).

"B" **Performance**: Best of the fitness values averaged as mean performance. (Represented as 'Best' in Table 2).

As can be observed COUNTER NICHING EA has demonstrated descent performance in majority of the test cases. However, as can be seen from the highlighted segment (*highlighted in bold*) of Table 2, the proposed algorithm was not very efficient in handling the comparatively higher dimensional cases (50 and 100 dimensional cases in this example) for the rotated Rastrigin Function $f_{rrtg}(x)$. Keeping in mind the concept of *No Free Lunch Theorem*, this is acceptable as no single algorithm can be expected to perform favorably for all possible test cases. The chosen benchmark test functions represent a wide variety of test cases.

An algorithm's value can only be established if its performance is tested against that of existing algorithms for similar purposes. In the next phase of our experiments we have presented comparative performances of COUNTER NICHING EA as against SEA, SOCEA, CEA, and DGEA.

6.2   Comparative Performance of COUNTER NICHING EA

Simulation results obtained with COUNTER NICHING EA in comparison to SEA, SOCEA, CEA, and DGEA (see Sect. 5.2 for descriptions of these algorithms) are

Table 2 Error values achieved on the test functions with simulation runs for COUNTER NICHING EA

		$f_{ack}(x)$	$f_{gri}(x)$	$f_{rtg}(x)$	$f_{ros}(x)$	$f_{elp}(x)$	$f_{sch-1.2}(x)$	$f_{rrtg}(x)$
20D	1st (Best)	1.00E-61	4.3E-62	1.11E-61	1.01E-60	2.01E-60	2.01E-50	3.89E-6
	7th	1.11E-61	4.41E-62	1.131E-61	1.01E-60	2.01E-60	2.01E-50	3.89E-6
	15th (Median)	1.11E-61	4.96E-62	1.210E-61	1.11E-60	2.89E-60	2.71E-50	3.9E-6
	22nd	1.95E-61	5.61E-62	2.02E-61	1.92E-60	2.91E-60	2.91E-50	3.93E-6
	30th (Worst)	3.11E-61	8.71E-62	3.30E-61	2.01E-60	2.91E-60	2.91E-50	3.99E-6
	Mean	1.12E-61	5.02E-62	1.21E-61	1.12E-60	2.92E-60	2.72E-50	3.9E-6
	Std.	8.33E-62	1.64E-62	8.6E-62	4.71E-61	4.62E-61	4.22E-51	3.88E-8
50D	1st (Best)	0.56E-29	1.00E-30	1.00E-30	1.21E-29	1.01E-30	2.21E-20	9.01
	7th	0.71E-29	1.01E-30	1.01E-30	1.41E-29	1.01E-30	2.40E-20	9.01
	15th (Median)	0.71E-29	1.01E-30	1.10E-30	1.90E-29	1.10E-30	2.90E-20	9.11
	22nd	0.91E-29	1.91E-30	1.81E-30	1.92E-29	1.51E-30	2.91E-20	9.22
	30th (Worst)	0.99E-29	1.99E-30	1.99E-30	1.98E-29	1.92E-30	2.91E-20	9.24
	Mean	0.73E-29	1.11E-30	1.11E-30	1.91E-29	1.11E-30	2.90E-20	9.12
	Std.	1.55E30	4.76E-31	4.41E-31	3.16E-30	3.66E-31	3.16E-21	0.098
100D	1st (Best)	1.00E-9	1.20E-9	1.90E-9	2.09E-9	2.09E-8	2.09E-5	10.52
	7th	1.01E-9	1.51E-9	1.92E-9	2.91E-9	2.92E-8	2.59E-5	10.66
	15th (Median)	1.12E-9	1.72E-9	1.99E-9	2.99E-9	2.99E-8	3.29E-5	11.09
	22nd	1.36E-9	1.86E-9	2.21E-9	3.21E-9	3.21E-8	3.79E-5	11.61
	30th (Worst)	1.36E-9	1.92E-9	2.92E-9	3.92E-9	3.90E-8	3.98E-5	11.79
	Mean	1.13E-9	1.81E-9	2.01E-9	3.03E-9	3.01E-8	3.69E-5	11.50
	Std.	1.61E-10	2.71E-10	3.88E-10	6.91E-10	5.81E-10	7.48E-6	0.5241

Dimensions of each function considered are 20, 50 and 100

Table 3 Average fitness comparison for SEA, SOCEA, CEA, DGEA, and COUNTER NICHING EA*

Function	SEA	SOCEA	CEA	DGEA	C_EA*
$f_{ack}(x)_{20D}$	2.494	0.633	0.239	3.36E-5	1.08E-61
$f_{gri}(x)_{20D}$	1.171	0.930	0.642	7.88E-8	4.6E-62
$f_{rtg}(x)_{20D}$	11.12	2.875	1.250	3.37E-8	1.21E-61
$f_{ros}(x)_{20D}$	8292.32	406.490	149.056	8.127	1.0E-60
$f_{elp}(x)_{20D}$	–	–	–	–	2.9E-60
$f_{sch-1.2}(x)_{20D}$	–	–	–	–	2.7E-50
$f_{rrtg}(x)_{20D}$	–	–	–	–	3.9E-6
$f_{ack}(x)_{50D}$	2.870	1.525	0.651	2.52E-4	1.01E-29
$f_{gri}(x)_{50D}$	1.616	1.147	1.032	1.19E-3	1.01E-30
$f_{rtg}(x)_{50D}$	44.674	22.460	14.224	1.97E-6	2.01E-30
$f_{ros}(x)_{50D}$	41425.674	4783.246	1160.078	59.789	1.91E-29
$f_{elp}(x)_{50D}$	–	–	–	–	1.00E-30
$f_{sch-1.2}(x)_{50D}$	–	–	–	–	2.9E-20
$f_{rrtg}(x)_{50D}$	–	–	–	–	9.1
$f_{ack}(x)_{100D}$	2.893	2.220	1.140	9.80E-4	1.00E-9
$f_{gri}(x)_{100D}$	2.250	1.629	1.179	3.24E-3	1.80E-9
$f_{rtg}(x)_{100D}$	106.212	86.364	58.380	6.56E-5	2.00E-9
$f_{ros}(x)_{100D}$	91251.300	30427.63	6053.870	880.324	3.00E-9
$f_{elp}(x)_{100D}$	–	–	–	–	2.99E-8
$f_{sch-1.2}(x)_{100D}$	–	–	–	–	3.7E-5
$f_{rrtg}(x)_{100D}$	–	–	–	–	11.51

Dimensions of each function considered are 20, 50 and 100. '–' appears where the corresponding data is not available

presented in Table 3. Results reported in this case, for COUNTER NICHING EA were averaged over 50 independent runs.

These simulation results demonstrate COUNTER NICHING EA's superior performance as regards to solution precision in all the test cases, particularly for lower dimensional instances. This may be attributed to COUNTER NICHING EA's ability to strike a better balance between exploration and exploitation. However, the proposed algorithm's performance deteriorates with increasing dimensions. Also, the algorithm could not handle the high dimensional versions of the high epistatis rotated Rastrigin function to any satisfactory level. Table 4 depicts the runtimes for the tested algorithms for the 100 dimensional scenarios of four test cases used in our experiments. Considering the structures of the algorithms, a trade-off between solution accuracy and computational time can be expected for COUNTER NICHING EA. On the other hand, DGEA, which is designed to skip certain genetic operations depending on the level of population diversity, would be a clear winner in terms of computation time if all the algorithms are executed for the same number of generations in each run.

Table 4 Average runtime in milliseconds for SEA, SOCEA, CEA, DGEA and COUNTER_NICHING_EA* for the 100 dimensional scenarios

Method	$f_{ack}(x)_{100D}$	$f_{gri}(x)_{100D}$	$f_{rtg}(x)_{100D}$	$f_{ros}(x)_{100D}$
SEA	1128405	1171301	1124925	1087615
SOCEA	1528864	1562931	1513691	1496164
CEA	2951963	3656724	2897793	2183283
DGEA	864316	969683	819691	883811
C_EA*	418489	521800	491411	510266

Average of 100 runs with 2000 generations for COUNTER_NICHING_EA* and 5000 generations for other algorithms

For the reported results as shown in Table 3, the 100 dimensional scenarios of the test problems used 5000 generations for each of the compared algorithm, namely, SEA, SOCEA, CEA and DGEA. On the other hand, COUNTER NICHING EA used only 2000 generations to reach the reported results. Hence, for comparison purposes it is only fair to consider the computation time required by the different methods to reach comparable results. As can be observed from Table 4, despite its relatively complex algorithmic structure, COUNTER NICHING EA requires less computation time to reach better or comparable solution accuracy. We have also extended the simulation runs beyond the fixed number of generations and to the stagnation point. Here, stagnation point is defined by the generation with 500 successive generations of no fitness improvement preceding it. Table 5 summarizes the results for DGEA and COUNTER NICHING EA with fixed run and at stagnation. Both DGEA and COUNTER NICHING EA show some improvement over the results obtained with fixed number of generations in most cases. COUNTER NICHING EA still outper-

Table 5 Average fitness comparison for DGEA and COUNTER_NICHING_EA*

Function	DGEA (Fixed run)	DGEA (Stagnation)	C_EA* (Fixed run)	C_EA* (Stagnation)
$f_{ack}(x)_{20D}$	8.05E-4	3.36e-5	1.08E-61	1.09E-62
$f_{ack}(x)_{50D}$	4.61E-3	2.52E-4	1.01E-29	1.01E-30
$f_{ack}(x)_{100D}$	0.01329	9.80E-4	1.00E-9	1.01E-10
$f_{gri}(x)_{20D}$	7.02E-4	7.88E-8	4.6E-62	4.01E-62
$f_{gri}(x)_{50D}$	4.40E-3	1.19E-3	1.01E-30	1.01E-31
$f_{gri}(x)_{100D}$	0.01238	3.24E-3	1.80E-9	1.52E-10
$f_{rtg}(x)_{20D}$	2.21E-5	3.37E-8	1.21E-61	1.00E-62
$f_{rtg}(x)_{50D}$	0.01664	1.97E-6	2.01E-30	2.01E-31
$f_{rtg}(x)_{100D}$	0.15665	6.56E-5	2.00E-9	2.00E-11
$f_{ros}(x)_{20D}$	96.007	8.127	1.0E-60	1.0E-60
$f_{ros}(x)_{50D}$	315.395	59.789	1.91E-29	1.90E-29
$f_{ros}(x)_{100D}$	1161.550	880.324	3.00E-9	3.00E-9

Dimension of each function in this case is 100. Both algorithms were executed till stagnation

forms DGEA. Also, COUNTER NICHING EA has arrived at these superior results in much fewer generations. However, no significant improvement was observed in case of all three different dimensional cases of the Rosenbrock function, in case of COUNTER NICHING EA.

6.3 An Analysis of Population Diversity for COUNTER NICHING EA

In the next phase of our experiments, we have investigated COUNTER NICHING EA's performance in terms of maintaining constructive diversity. There are various measures of diversity available. The "*distance-to-average-point*" measure used in [9] is relatively robust with respect to population size, dimensionality of problem and the search range of each variable. Hence, we have used this measure of diversity in our investigation. The "*distance-to-average-point*" measure for N dimensional numerical problems can be described as below [9].

$$diversity\,(P) = \frac{1}{|L| \cdot |P|} \cdot \sum_{i=1}^{|P|} \sqrt{\sum_{j=1}^{N} \left(s_{ij} - \bar{s}_j\right)^2} \qquad (1)$$

where, $|L|$ is the length of the diagonal or range in the search space $S \subseteq \Re^N$, P is the population, $|P|$ is the population size, N is the dimensionality of the problem, s_{ij} is the j'th value of the i'th individual, and $\bar{s}_j$ is the j'th value of the average point $\bar{s}$. It is assumed that each search variable s_k is in a finite range, $s_{k\_min} \le s_k \le s_{k\_max}$. Table 6 depicts the average diversity for four test problems with COUNTER NICHING EA simulation runs. The values reported in Table 6, averages the value of the diversity measure in Eq. (1) calculated at each generation where there has been an improvement in average fitness over 500, 1000 and 2000 generations for the 20, 50 and 100 dimensional cases respectively. Final values were averaged over 100 runs. To eliminate the noise in the initial generations of a run, diversity calculation does not start until the generation since which a relatively steady improvement in fitness has been observed. Table 6 shows that the COUNTER NICHING EA does not necessarily maintain very high average population diversity. However, EA's requirement is not

Table 6 Average population diversity comparison for COUNTER NICHING EA (fixed run)

Function	20D	50D	100D
$f_{ack}\,(x)$	0.001350	0.001811	0.002001
$f_{gri}\,(x)$	0.001290	0.001725	0.002099
$f_{rtg}\,(x)$	0.003000	0.003550	0.004015
$f_{ros}\,(x)$	0.001718	0.002025	0.002989

An average of 100 runs have been reported in each case

Table 7 The *P*-values of the t-test with 99 degrees of freedom

Function	C_EA*–SEA	C_EA*–SOCEA	C_EA*–CEA	C_EA*–DGEA
$f_{ack}(x)_{20D}$	0.1144	0.4263	0.625	0.9954
$f_{gri}(x)_{20D}$	0.2793	0.3349	0.4231	0.9998
$f_{rtg}(x)_{20D}$	0.0009	0.0901	0.2636	0.9999
$f_{ros}(x)_{20D}$	0	0	0	0.0044
$f_{ack}(x)_{50D}$	0.0903	0.217	0.4198	0.9873
$f_{gri}(x)_{50D}$	0.2037	0.2843	0.3098	0.9725
$f_{rtg}(x)_{50D}$	0	0	0.0002	0.9989
$f_{ros}(x)_{50D}$	0	0	0	0
$f_{ack}(x)_{100D}$	0.0891	0.1363	0.2857	0.975
$f_{gri}(x)_{100D}$	0.1337	0.2019	0.2776	0.9546
$f_{rtg}(x)_{100D}$	0	0	0	0
$f_{ros}(x)_{100D}$	0	0	0	0

Dimensions of each function considered are 20, 50 and 100. '–' appears where the corresponding data is not available

to maintain very high average population diversity but to maintain an optimal level of population diversity. The high solution accuracy obtained by COUNTER NICHING EA proves that the algorithm is successful in this respect.

6.4 Statistical Significance of Comparative Analysis

Finally, a *t*-test (at 0.05 level of significance; 95 % confidence) was applied in order to ascertain if differences in the "*A*" performance for the best average fitness function are statistically significant from the other techniques used for comparison. The **P**-values of the two-tailed *t*-test are given in Table 7. As can be observed, the difference in "*A*" performance of COUNTER NICHING EA is statistically significant compared to the majority of the techniques across the test functions in their three different dimensional versions.

7 Conclusions

In this paper we investigated the issues related to population diversity in the context of the evolutionary search process. We established the association between population diversity and the search ability of a typical evolutionary algorithm. Then we presented an investigation on an intelligent mutation based EA that tries to achieve optimal diversity in the search landscape. The framework basically incorporates two key processes. *Firstly*, the population's spatial information is obtained with a pseudo-niching algorithm. *Secondly*, the information is used to identify potential local convergence and community formations. Then diversity is introduced with informed genetic operations, aiming at two objectives: (a) Promising samples from unexplored regions are introduced replacing *redundant* less fit members of over-populated communities and (b) while local entrapment is discouraged, representative members are still preserved to encourage *exploitation*. While the current focus of the research was to introduce and maintain population diversity to avoid local entrapment, this Counter Niching-based algorithm can also be adapted to serve as an inexpensive alternative for *niching* genetic algorithm, to identify multiple solutions in multimodal problems as well as to suit the diversity requirements in a dynamic environment.

References

1. Bhattacharya, M.: An informed operator approach to tackle diversity constraints in evolutionary search. In: Proceedings of The International Conference on Information Technology, ITCC 2004, vol. 2, pp. 326–330. IEEE Computer Society Press. ISBN 0-7695-2108-8
2. Bhattacharya, M.: Counter-niching for constructive population diversity. In: Proceedings of the 2008 IEEE Congress on Evolutionary Computation (CEC 2008), pp. 4174–4179. IEEE Press, Hong Kong. ISBN: 978-1-4244-1823-7
3. Friedrich, T., Oliveto, P.S., Sudholt, D., Witt, C.: Theoretical analysis of diversity mechanisms for global exploration. In: Proceedings of the Genetic and Evolutionary Computation Conference, pp. 945–952 (2008)
4. Friedrich, T., Hebbinghaus, N., Neumann, F.: Rigorous analyses of simple diversity mechanisms. In: Proceedings of the Genetic and Evolutionary Computation Conference, pp. 1219–1225 (2007)
5. Ganv'an-L'opez, E., McDermott, J., O'Neill, M., Brabazon, A.: Towards an understanding of locality in genetic programming. In: Proceedings of the 12th Annual Conference on Genetic and Evolutionary Computation, pp. 901–908 (2010)
6. De Jong, K.A.: An analysis of the behavior of a class of genetic adaptive systems. PhD thesis, University of Michigan, Ann Arbor, MI, Dissertation Abstracts International 36(10), 5140B, University Microfilms Number 76–9381 (1975)
7. Leung, Y., Gao, Y., Xu, Z.B.: Degree of population diversity-a perspective on premature convergence in genetic algorithms and its Markov chain analysis. IEEE Trans. Neural Netw. **8**(5), 1165–1176 (1997)
8. Liang, Y., Leung, K.S.: Genetic algorithm with adaptive elitist-population strategies for multimodal function optimization. Appl. Soft Comput. **11**(2), 2017–2034 (2011)
9. Ursem, R.K.: Diversity-guided evolutionary algorithms. In: Proceedings of Parallel Problem Solving from Nature VII (PPSN-2002), pp. 462–471 (2002)
10. Thomsen, R., Rickers, P.: Introducing spatial agent-based models and self-organised criticality to evolutionary algorithms. Master's thesis, University of Aarhus, Denmark (2000)
11. Bäck, T., Fogel, D.B., Michalewicz, Z., et al. (eds.): Handbook on Evolutionary Computation. IOP Publishing Ltd and Oxford University Press (1997)
12. Bhattacharya, M., Nath, B.: Genetic programming: a review of some concerns. In: Computational Science-ICCS 2001, pp. 1031–1040. Springer, Heidelberg (2001)
13. Adra, S.F., Fleming, P.J.: Diversity management in evolutionary many-objective optimization. IEEE Trans. Evol. Comput. **15**(2), 183–195 (2011)
14. Araujo, L., Merelo, J.J.: Diversity through multiculturality: assessing migrant choice policies in an island model. IEEE Trans. Evol. Comput. **15**(4), 456–468 (2011)

15. Chow, C.K., Yuen, S.Y.: An evolutionary algorithm that makes decision based on the entire previous search history. IEEE Trans. Evol. Comput. **15**(6), 741–769 (2011)
16. Curran, D., O'Riordan, C.: Increasing population diversity through cultural learning. Adapt. Behav. **14**(4), 315–338 (2006)
17. Gao, H., Xu, W.: Particle swarm algorithm with hybrid mutation strategy. Appl. Soft Comput. **11**(8), 5129–5142 (2011)
18. Jia, D., Zheng, G., Khan, M.K.: An effective memetic differential evolution algorithm based on chaotic local search. Inf. Sci. **181**(15), 3175–3187 (2011)
19. Bhattacharya, M.: Meta model based EA for complex optimization. Int. J. Comput. Intell. **4**, 1 (2008)
20. Bhattacharya, M.: Surrogate based EA for expensive optimization problems. In: IEEE Congress on Evolutionary Computation (2007)
21. Bhattacharya, M.: Reduced computation for evolutionary optimization in noisy environment. In: Proceedings of the 10th annual Conference Companion on Genetic and Evolutionary Computation. ACM (2008)
22. Bhattacharya, M.: Expensive optimization, uncertain environment: an EA-based solution. In: Proceedings of the 2007 GECCO conference companion on Genetic and evolutionary computation. ACM (2007)
23. Bhattacharya, M.: Meta model based EA for complex optimization. Int. J. Comput. Intell. **4**, 1 (2008)
24. Bhattacharya, M.: Exploiting landscape information to avoid premature convergence in evolutionary search. In: IEEE Congress on Evolutionary Computation (2006)
25. Ishibuchi, H., Narukawa, K., Tsukamoto, N., Nojima, Y.: An empirical study on similarity-based mating for evolutionary multi-objective combinatorial optimization. Eur. J. Oper. Res. **188**(1), 57–75 (2008)

Probabilistic Planning in AgentSpeak Using the POMDP Framework

Kim Bauters, Kevin McAreavey, Jun Hong, Yingke Chen, Weiru Liu,
Lluís Godo and Carles Sierra

Abstract AgentSpeak is a logic-based programming language, based on the Belief-Desire-Intention paradigm, suitable for building complex agent-based systems. To limit the computational complexity, agents in AgentSpeak rely on a plan library to reduce the planning problem to the much simpler problem of plan selection. However, such a plan library is often inadequate when an agent is situated in an uncertain environment. In this work, we propose the AgentSpeak$^+$ framework, which extends AgentSpeak with a mechanism for probabilistic planning. The beliefs of an AgentSpeak$^+$ agent are represented using epistemic states to allow an agent to reason about its uncertain observations and the uncertain effects of its actions. Each epistemic state consists of a POMDP, used to encode the agent's knowledge of the environment, and its associated probability distribution (or belief state). In addition, the POMDP is used to select the optimal actions for achieving a given goal, even when faced with uncertainty.

1 Introduction

Using the Belief-Desire-Intention (BDI) agent architecture [20], we can develop complex systems by treating the various system components as autonomous and interactive agents [12]. The beliefs determine the desires that are achievable, the desires are the goals an agent wants to achieve and the intentions are those desires the agent is acting upon. A number of successful agent-oriented programming languages have been developed based on this architecture, such as AgentSpeak [19] and CAN [21]. Notable BDI implementations include, for example, JASON [5] and JADEX [6]. The benefits of the BDI model in scalability, autonomy and intelligence have been illustrated in various application domains such as power engineering [15] and control

K. Bauters (✉) · K. McAreavey · J. Hong · Y. Chen · W. Liu · L. Godo · C. Sierra
Queen's University Belfast (QUB), Belfast, UK
e-mail: k.bauters@qub.ac.uk

L. Godo · C. Sierra
IIIA, CSIC, Bellaterra, Spain

© Springer International Publishing Switzerland 2016
I. Hatzilygeroudis et al. (eds.), *Combinations of Intelligent Methods
and Applications*, Smart Innovation, Systems and Technologies 46,
DOI 10.1007/978-3-319-26860-6_2

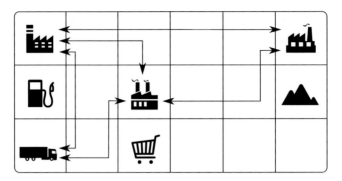

Fig. 1 The material collection scenario

systems [12]. Key to the efficiency of BDI agents is the use of a set of pre-defined plans, which simplify the planning problem to an easier plan selection problem. However, obtaining a plan library that can cope with every possible situation requires adequate domain knowledge. This knowledge is not always available, particularly when dealing with uncertain situations. As such, when faced with uncertainty, an autonomous and intelligent agent should resort to other forms of planning to make rational decisions.

To illustrate the problem, consider the example shown in Fig. 1. A truck needs to collect materials from three different factories, each producing a distinct type of material that may or may not be available (i.e. the environment is *stochastic*). The truck needs to collect all materials by visiting each factory while limiting costs (e.g. fuel). The truck agent is uncertain as to whether the material in a factory is ready to collect, but it can use previous experience to estimate a degree of belief. To further complicate the situation, the truck agent can only infer its location by observing nearby signposts (e.g. the agent is near a supermarket or a petrol station). Travelling between factories may also fail (i.e. non-deterministic actions).

The large number of possibilities make a pre-defined plan library infeasible, even in this small example. We address these issues by combining AgentSpeak with Partially Observable Markov Decision Processes (POMDPs). POMDPs are a framework for probabilistic planning [13], and are often used as a decision theory model for agent decision making. Other frameworks, such as probabilistic graphplan [4], only consider the *uncertain effects* of actions. Similarly, the *partial observability* of the *stochastic* environment is not addressed by approaches such as probabilistic Hierarchical Task Networks (HTN) [17] and Markov Decision Processes (MDPs) [3]. As such, POMDPs seem to offer an elegant solution to deal with examples such as the one discussed above. In particular, when optimal solutions are required (e.g. the truck wants to collect as many materials as possible subject to the fuel limit), POMDPs can be used to compute these solutions. However, even though efficient algorithms to compute the optimal policy exist (e.g. [11]), POMDPs are still computationally expensive.

By integrating POMDPs into a BDI architecture, we retain the scalability of the BDI architecture, while adding to it the ability to model uncertainty as well as on-demand access to optimal actions (provided by the POMDP component). The framework we propose is called AgentSpeak$^+$. In this framework we introduce the concept of epistemic states [14] to the BDI paradigm. These epistemic states are used to model the beliefs of an agent about uncertain information, along with information on how these beliefs evolve over time. To achieve this, POMDPs are embedded into the agent's epistemic states and are used to represent some aspects of the agent's domain knowledge. The optimal actions generated by these POMDPs can be fed into the agent's plan execution. Therefore, alongside the traditional trigger-response mechanism based on pre-defined plans in BDI, an AgentSpeak$^+$ agent also has the ability to take optimal actions when dealing with uncertain and partially observable environments.

The main contributions of this work are as follows. First, we extend the belief base of a BDI agent with epistemic states consisting of POMDPs to allow an agent to reason about both the partially observable stochastic environment and the uncertain effects of its actions. Second, we present how the agent can delegate POMDPs to find optimal action(s) when the agent is dynamically generating its plans under uncertainty for achieving its goals. Finally, we demonstrate through a scenario discussion how the proposed framework can be used to design agents that are aware of the uncertainty in the environment and are able to react accordingly.

The remainder of our work is organised as follows. Preliminary notions on AgentSpeak and POMDPs are mentioned in Sect. 2. In Sect. 3 we propose the AgentSpeak$^+$ architecture which integrates POMDPs into AgentSpeak. A scenario is discussed in Sect. 4. Related work is discussed in Sect. 5 and in Sect. 6 we conclude our work.

2 Preliminaries

We start with some preliminaries on AgentSpeak (see Sect. 2.1) and Partially Observable Markov Decision Processes (POMDP) (see Sect. 2.2).

2.1 AgentSpeak

We first define how an agent program can be written. We use S to denote a finite set of symbols for predicates, actions, and constants, and V to denote a set of variables. Following convention in logic programming, elements from S are written using lowercase letters and elements from V using uppercase letters. We use the standard first-order logic definition of a term[1] and we use **t** as a compact notation for $t_1, \ldots, t_n$, i.e. a vector of terms. We have [19]:

Definition 1 If b is a predicate symbol, and **t** are terms, then $b(\mathbf{t})$ is a *belief atom*. If $b(\mathbf{t})$ and $c(\mathbf{s})$ are belief atoms, then $b(\mathbf{t})$, $\neg b(\mathbf{t})$, and $b(\mathbf{t}) \wedge c(\mathbf{s})$ are *beliefs*.

Definition 2 If $g(\mathbf{t})$ is a belief atom, then $!g(\mathbf{t})$ and $?g(\mathbf{t})$ are goals with $!g(\mathbf{t})$ an *achievement goal* and $?g(\mathbf{t})$ a *test goal*.

Definition 3 If $p(\mathbf{t})$ is a belief atom or goal, then $+p(\mathbf{t})$ and $-p(\mathbf{t})$ are *triggering events* with $+$ and $-$ denoting the addition and deletion of a belief/goal, respectively.

Definition 4 If a is an action symbol and **t** are terms, then $a(\mathbf{t})$ is an *action*.

Definition 5 If e is a triggering event, $h_1, \ldots, h_m$ are beliefs and $q_1, \ldots, q_n$ are goals or actions, then $e : h_1 \wedge \ldots \wedge h_m \leftarrow q_1, \ldots, q_n$ is a *plan*. We refer to $h_1 \wedge \ldots \wedge h_m$ as the *context* of the plan and to $q_1, \ldots, q_n$ as the *plan body*.

Following these definitions, we can now specify an agent by its belief base **BB**, its plan library **PLib** and the action set **Act**. The belief base of an agent, **BB**, which is treated as a set of belief atoms, contains the information that the agent has about the environment. The plan library contains those plans that describe how the agent can react to the environment, where plans are triggered by events. Finally, the action set simply describes the primitive actions to which the agent has access.

On the semantic level, the state of an AgentSpeak agent $\mathbb{A}$ can be described by a tuple $\langle$**BB, PLib, E, Act, I**$\rangle$, with **E** the event set, **I** the intention set and **BB, Plib** and **Act** as before [19]. The event set and intention set are mainly relevant during the execution of an agent. Intuitively, the event set contains those events that the agent still has to deal with. When an agent reacts to one of these new events e, it selects those plans that have e as the triggering event, i.e. the *relevant* plans. We say that a plan is *applicable* when the context of the plan evaluates to true according to the belief base of the agent. For a given event e there may be many applicable plans, one of which is selected and added to an intention. The intention set is a set of intentions that are currently being executed concurrently, i.e. those desires that the agent has chosen to pursue. Each intention is a stack of partially executed plans and is itself executed by performing the actions in the body of each plan in the stack. The execution of an intention may change the environment and/or the agent's beliefs. Also, if the execution of an intention results in the generation of new internal events (or subgoals), then additional plans may be added to the stack.

---

[1]A variable is a term, a constant is a term, and from every n terms $t_1, t_2, \ldots, t_n$ and every n-ary predicate p a new term $p(t_1, t_2, \ldots, t_n)$ can be created.

2.2 POMDP

Partially Observable Markov Decision Processes (POMDPs) have gained popularity as a computational model for solving probabilistic planning problems in a *partially observable* and *stochastic* environment (see e.g. [13]). They are defined as follows:

Definition 6 A POMDP $\mathcal{M}$ is a tuple $\mathcal{M} = \langle S, A, \Omega, R, T, O \rangle$ where S, A, and Ω are sets of states, actions, and observations, respectively. Furthermore, $R : S \times A \to \mathbb{R}$ is the reward function, $T : S \times A \to \Delta(S)$ is the transition function and $O : S \times A \to \Delta(\Omega)$ is the observation function. Here, $\Delta(\cdot)$ is the space of probability distributions.

Instead of knowing the current state exactly, there is a probability distribution over the state space S, called the *belief state*[2] $b(S)$, with $b(s)$ the probability that the current state is s. When S is clear from the context, we simply write b instead of $b(S)$. In Definition 6, we have that the *Markovian* assumption is encoded in the transition function since the new state depends only on the previous state. Given the belief state b_t at time t, after performing action a and receiving observation o, the new belief state b_{t+1} is obtained using the Bayesian rule:

$$
\begin{aligned}
b_{t+1}(s) &= P(s \mid b_t, a, o) \\
&= \frac{O(o \mid s, a) \cdot \sum_{s' \in S} T(s \mid s', a) \cdot b_t(s')}{P(o \mid b_t, a)}
\end{aligned}
\tag{1}
$$

where $P(o \mid b_t, a)$ is a normalisation factor obtained by marginalising s out as follows:

$$
P(o \mid b_t, a) = \sum_{s \in S} O(o \mid s, a) \cdot \sum_{s' \in S} T(s \mid s', a) \cdot b_t(s').
$$

The decision a at horizon t takes into account both the instant reward as well as all possible rewards in future decision horizons (of the POMDP execution). Given a POMDP $\mathcal{M}$, its policy $\pi : \mathfrak{B} \to A$ is a function from the space of belief states (denoted as $\mathfrak{B}$) to the set of actions. The policy provides the optimal action to perform for a given belief state at each decision horizon, i.e. it is the action that should be performed in the current belief state to maximise the expected reward.

Definition 7 (*Optimal Policy*) Given a POMDP $\mathcal{M}$ with the initial belief state b_1, π^* is an optimal policy over the next H decision horizons if it yields the highest cumulated expected reward value V^*:

$$
V^*(b_1) = \sum_{t=1}^{H} \gamma^{t-1} \cdot R(s, \pi^*(b_t)) \cdot b_t(s)
$$

---

[2]Not to be confused with the *belief base* of an agent, which we see later.

where $b_t(s)$ is updated according to Eq. (1) and $\gamma \in (0, 1]$ is a discounting factor to ensure that future rewards are lower. Here $\pi^*(b_t)$ is the action determined by policy π^* and the belief state b_t.

A probabilistic planning problem is then defined as the problem of finding the optimal actions for a given POMDP $\mathcal{M}$ and an initial belief state b (where the optimal actions in a POMDP setting are described using a policy π).

3 Integration of AgentSpeak and POMDPs

We now discuss how AgentSpeak and POMDP can be integrated into a single framework. The resulting framework, called AgentSpeak$^+$, allows us to define agents that can perform on-demand planning based on the POMDP to provide optimal decisions in an uncertain environment. We start by introducing the concept of epistemic states to model the uncertain beliefs of an agent. We define an epistemic state in Sect. 3.1 as containing both a POMDP $\mathcal{M}$ and a belief state b, where the former encodes the agent's domain knowledge about the partially observable environment and the latter represents the current uncertain information about the states modelled in $\mathcal{M}$. The basic concepts needed for probabilistic planning are introduced in Sect. 3.2, where we show how a new construct behaving as an action in AgentSpeak allows for the desired on-demand planning. Our new framework, AgentSpeak+ in then introduced in Sect. 3.3, where it combines both aforementioned ideas with classical AgentSpeak.

3.1 Epistemic States

Normally, a belief base only contains belief atoms with Boolean values. Such an approach is insufficient to reason over the uncertain beliefs of the agent. To overcome this, we extend the the idea of a belief base into the concept of an epistemic state.

Definition 8 (*Epistemic states*) Let $\mathcal{M}$ be a POMDP which models the situated partially observable stochastic environment. By definition, $\mathcal{M}$ includes a set of states S. The epistemic state Φ over the state space S is defined as $\Phi = \langle b, \mathcal{M} \rangle$, where $b : S \to [0, 1]$ is a probability distribution over S.

The POMDP $\mathcal{M}$ defined in the epistemic state Φ represents the knowledge about the uncertain environment. The observations Ω and state space S in $\mathcal{M}$ are subject to the agent $\mathbb{A}$'s belief base. The action set A of the POMDP can contain both primitive actions in agent $\mathbb{A}$'s action set **Act** and *compound* actions which correspond to goals achievable by executing existing plans. *Compound* actions, which are plans in BDI, can have various outcomes. However, they can be transformed into primitive actions over which POMDP can reason using a translation such as the one proposed in

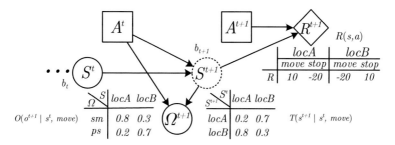

Fig. 2 Graphical representation of the POMDP $\mathcal{M}$. We have that $S = \{\mathsf{LocA}, \mathsf{LocB}\}$, $\Omega = \{\mathsf{(s)uper(m)arket}, \mathsf{(p)etrol\ (s)tation}\}$ and $A = \{\mathsf{move}, \mathsf{stop}\}$. The transition, observation and reward functions are shown as tables

[9]. Indeed, the work in [9] suggests an approach to summarise sceptical results of compound actions as primitive actions. The belief state b quantitatively denotes the degree of certainty about the partial state space as the initial condition for $\mathcal{M}$.

Example 1 Let $\Phi = \langle b, \mathcal{M} \rangle$ be an epistemic state. The state space S in $\mathcal{M}$ is $\{\mathsf{LocA}, \mathsf{LocB}\}$, i.e. two possible locations, where we want to be in LocB (as indicated by the reward function). The current belief state is given by $b(\mathsf{LocA}) = 0.6$ and $b(\mathsf{LocB}) = 0.4$, which will change based on the actions performed by the agent (e.g. probably in LocB if you move from LocA) and its observations (e.g. probably in LocA when you see a supermarket). The POMDP $\mathcal{M}$ encoding the relevant knowledge is graphically illustrated in Fig. 2.

Each epistemic state contains all knowledge needed to plan over and reason about a well-defined subset of the environment. Given that each epistemic state has an optimal policy, this optimal policy intuitively encodes a subplan for this subset of the environment. Throughout this work, we assume that a corresponding symbol Φ is available on the syntactic level as well, i.e. on the level of an AgentSpeak agent we are able to refer to a specific epistemic state. An executed action and a newly obtained observation are together taken as new input to revise the epistemic state using the belief updating process in Eq. (1), where revision is defined as:

Definition 9 (*Epistemic state revision*) Let $\Phi = \langle b, \mathcal{M} \rangle$ be an epistemic state and $I = \langle a, o \rangle$ an input with $a \in A$ an action and $o \in \Omega$ an observation. The revision of Φ by I, denoted as $\Phi \circ I$, is defined as:

$$\Phi \circ I = \langle b', \mathcal{M} \rangle$$

with $\circ$ a revision operator. Particularly, $\circ$ is given by Eq. (1). The result of revision is a new epistemic state, where the new belief state b' is determined based on the old belief state b and the input I.

Example 2 (*Example 1 continued*) Let $\Phi = \langle b, \mathcal{M} \rangle$ be the epistemic state from Example 1 with the belief state b. After performing action **move**, and receiving a new observation **ps**, the revised probability distribution over possible locations is $b'(\mathsf{LocA}) = 0.16$ and $b'(\mathsf{LocB}) = 0.84$.

It is important to note that revision will only revise the belief state, while keeping the corresponding POMDP unchanged (i.e. revision does not alter the domain knowledge of the agent in this case). When there is a sequence of inputs $I_1, \ldots, I_n$ the epistemic state is simply revised iteratively. Furthermore, we assume that an agent can have multiple epistemic states, each dealing with a localised and isolated part of the beliefs of the agent. For example, the availability of material at each factory is independent of the colour of the traffic light. Localised epistemic states allow an agent to revise a corresponding epistemic state given a new input without affecting other epistemic states.[3] This reflects, to some extent, the notion of *minimal change principle* in belief revision.

We will also allow the belief state (i.e. the beliefs maintained by a POMDP) to be extracted from the agent's *belief base* (i.e. the component of an AgentSpeak agent where beliefs are stored). This is useful for designing an AgentSpeak$^+$ agent, as it allows the automatic extraction of the belief state of the POMDP from the AgentSpeak$^+$ program. To simplify the explanation, we will explicitly add the POMDP as a parameter to the belief atoms to make explicit to which POMDP the belief atom is associated.

Definition 10 (*Correlated belief atoms*) Two belief atoms $h(x, m, \mathcal{M})$ and $h(x', m', \mathcal{M})$ are said to be correlated if x and x' are two states of variable X_j (which is one of the variables in the joint state space S) defined in the POMDP $\mathcal{M}$, where m and m' are their corresponding probability values.

Definition 11 (*Extraction*) Let $\{h(x_i, m_i, \mathcal{M}) \mid i \in 1, \ldots, k\}$ be a set of exhaustively correlated belief atoms for variable X_j. The belief state $b(x_i) = m_i$ can be directly derived from this set iff $\sum_{i=1}^{k} m_i = 1$ and $X_j = \{x_1, \ldots, x_k\}$.

Here, a set of exhaustively correlated belief atoms implies that no other belief atoms in the agent belief base are correlated to any of the belief atoms in this set.

When the state space S of a POMDP $\mathcal{M}$ has a set of variables $\{X_1, \ldots, X_n\}$, then the belief state $b(S)$ is the joint probability distribution obtained from $b(X_j)$. When the belief state cannot be extracted from an agent's initial beliefs we assume a default probability distribution, i.e. a uniform distribution, for the belief state. Finally, whenever an epistemic state is initiated or revised the belief base of the agent will be updated accordingly using the corresponding triggering events $-h(x_i, m_i, \mathcal{M})$ and $+h(x_i, m'_i, \mathcal{M})$ in AgentSpeak.

---

[3]For simplicity, we restrict ourselves in this work to the case where each input is relevant to only one epistemic state.

We can also derive ordinary beliefs from the belief state:

Definition 12 (*Derivation*) Let $\Phi = \langle b, \mathcal{M} \rangle$ be an epistemic state containing a probability distribution over S. The belief atom of Φ, denoted as $Bel(\Phi)$, is derived as

$$Bel(\Phi) = \begin{cases} s_i, & when \quad P(S = s_i) \geq \delta \\ \mathsf{T}, & otherwise \end{cases}$$

Here δ is a pre-defined threshold for accepting that s_i represents the real world concerning S. Notation T is a special constant representing an agent's *ignorance*, i.e. an agent is not certain about the state of variable S.

Example 3 (*Example 2 continued*) The belief state b can be modelled as the belief atoms location(LocA, 0.6, $\mathcal{M}$) and location(LocB, 0.4, $\mathcal{M}$). The revised belief state b' is represented as location(LocA, 0.16, $\mathcal{M}$) and location(LocB, 0.84, $\mathcal{M}$).

3.2 Probabilistic Planning

Now that we have defined an epistemic state (which can deal with uncertainty) and how to revise it, we look at how probabilistic planning can be integrated into AgentSpeak. The POMDPs we use in the epistemic state allow us to decide the optimal action at each decision horizon by taking into account the immediate expected reward and the future rewards. However, simply computing the optimal plan at each step would severely hamper the reactiveness of the AgentSpeak agent due to the computational cost. Instead, we introduce a new action to AgentSpeak, ProbPlan, which can be used in AgentSpeak plans to explicitly compute the optimal action to achieve a goal for a given epistemic state $\mathcal{M}$. This enables the agent to react optimally when needed, e.g. for when performing the wrong action likely carries a high penalty. When optimality is not required or when reactiveness is of primary importance, the agent can instead rely on the abstractness and high performance of the normal BDI plan selection strategy.

Definition 13 (*Probabilistic planning action*) Let ProbPlan be an ordinary AgentSpeak action symbol and $\Phi = \langle b, \mathcal{M} \rangle$ an epistemic state. We say that ProbPlan(Φ, H) is a *probabilistic planning action*, with H the number of steps and corresponding rewards we should consider (i.e. H is the horizon). The effect of executing ProbPlan(Φ, H) is that the probabilistic planning problem defined by a POMDP $\mathcal{M}$ with initial belief state b and horizon H is solved, after which the optimal action $a_i \in A$ is executed.

Importantly, the action set A defined in a POMDP can contain both primitive actions and *compound* actions (i.e. subgoals), each representing different levels of planning granularity. In the latter case, a pre-defined plan in AgentSpeak will be triggered to pursue the goal corresponding to the optimal action a_i. This allows the optimal plan to be as specific as possible (to reach the goal without taking excess steps) while being as abstract as possible (to fully take effect of the domain knowledge encoded in the set of pre-defined plans). The effects of these compound actions can be computed either sceptically (i.e. only considering effects shared by all relevant plans) or credulously, which allows us to balance optimality and reactiveness for the given problem. In the first case, we guarantee the outcome of those effects that we want to bring about, but we leave it up to the AgentSpeak reasoning system to select the best plan at time of execution (i.e. we are not interested in the side-effects). In the latter case, we apply a *"best effort"* strategy, where we lose some optimality but gain reactiveness. In addition, it should be noted that while the result of ProbPlan(Φ, H) is an optimal action at the time of computation, there is no guarantee that this action will still be optimal during execution. Indeed, the optimal action/subgoal is not (by default) immediately executed and may be intertwined with the execution of other subgoals which alter the environment. The benefit of not enforcing this optimality but rather *trying* to be optimal is that we retain the reactiveness of BDI and are able to fully use the knowledge already encoded by the system developer in the subgoals.

For the running example, we consider a POMDP with the state space defined as $\{O, A, B, C\}$, i.e. the origin location O and three factories A, B and C. In addition, there is an action set consisting of 9 subgoals: 3 subgoals to go from the origin to a factory; and 6 subgoals to go from one factory to another. In all cases, the subgoals consist of both going to the location as well as collecting the corresponding material. For example, we will use goOBcollect to denote the subgoal to move from the origin to factory B in order to collect material B.

Definition 14 (*Probabilistic planning plan*) A plan pl is called a *probabilistic planning plan* if it contains at least one probabilistic planning action ProbPlan in the plan body.

Similar to Definition 13, a probabilistic planning plan is still a normal AgentSpeak plan. Due to the fact that each probabilistic planning problem defined on $\mathcal{M}$ always has a (not necessarily unique) optimal action, a probabilistic planning plan does not introduce infinite recursion. Furthermore, an optimisation can be applied to reduce the computational cost. Indeed, whenever the first optimal action is decided, a complete optimal policy over H decision horizons has already been constructed as part of the probabilistic planning problem. Before deliberating over the next action, the epistemic state will be revised with the current action and the new observation as defined in Definition 9. Given the revised epistemic state, the next optimal action can then be decided instantly based on the optimal policy that is already constructed without requiring extra computation.

Example 4 Consider the truck agent $\mathbb{A}_t$ from the running example where Φ is the relevant epistemic state. We have:

```
P1: +!collectMaterial : true ← ProbPlan(Φ, 3);
                                ProbPlan(Φ, 2);
                                ProbPlan(Φ, 1).
P2: +!goOAcollect : true ← moveOtoS1; !waitS1toA; moveS1toA;
                           senseLocation; !load(a).
```

The first plan describes how the truck agent can collect all the materials, i.e. how it can achieve its goal !collectMaterial. Due to some hard constraint (e.g. we only have limited fuel) we rely on the POMDP planning to plan ahead and figure out the best course of action. Since the abstract level considered by the POMDP in the epistemic state Φ can move from one factory to another in a single step, we consider a decision horizon of 3. The result of ProbPlan(Φ, 3) can for example be the subgoal goOAcollect, i.e. given all the information available to POMDP at the moment, the optimal action is to first visit factory A. During the execution of this subgoal new observations will be collected (e.g. through senseLocation) and will be taken into account when deliberating over ProbPlan(Φ, 2) by (implicitly) using the revised epistemic state to find the optimal action/subgoal to collect the remaining two materials.

3.3 AgentSpeak$^+$

We are now ready to define our AgentSpeak$^+$ framework:

Definition 15 (*AgentSpeak$^+$ agent*) An AgentSpeak$^+$ agent $\mathbb{A}^+$ is defined as a tuple $\langle$BB$^+$, EpS, PLib$^+$, E, Act, I$\rangle$, where the belief base BB$^+$ now contains belief atoms with an associated probability value, EpS is a set of epistemic states, the plan library PLib$^+$ contains an additional set of probabilistic planning plans, and E, Act and I are as before.

Normally, in AgentSpeak, the context of a plan consists of classical belief atoms. However, in AgentSpeak$^+$ the belief base contains uncertain belief atoms, i.e. we need a way to determine if a given context is *sufficiently plausible*.

Definition 16 (*Belief entailment*) Let $\{h(x_i, m_i, \mathcal{M}) \mid i \in 1, \ldots, k\}$ be a set of exhaustively correlated belief atoms in an agent belief base. The belief atom $h'(x_i)$ is entailed by the agent's belief base BB iff there exists $h(x_i, m_i, \mathcal{M}) \in$ BB such that $m_i \geq \delta$ with $0 < \delta \leq 1$. The value δ is context-dependent, and reflects the degree of uncertainty we are willing to tolerate.

Example 5 (*Example 3 continued*) The revised belief base contains the belief atoms location(LocA, 0.16, $\mathcal{M}$) and location(LocB, 0.84, $\mathcal{M}$). Given a threshold $\delta_{loc} = 0.8$, only the belief atom location(LocB) is entailed.

Notice that we can straightforwardly represent classical belief atoms by associating a probability of 1 with them. Verifying if a context is entailed, *i.e.* a conjunction of belief literals, is done classically based on the entailed belief atoms. As such, we recover classical AgentSpeak entailment of contexts if we enforce that belief entailment is only possible when $\delta = 1$. The revised reasoning cycle of AgentSpeak$^+$ agent is shown in Fig. 3. The agent now contains a set of epistemic states, each of which includes a POMDP. A new input can either revise the belief state of a POMDP or be inserted into the agent's belief base BB (i.e. this happens when the input is not

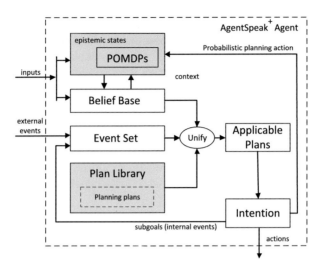

Fig. 3 The revised reasoning cycle for an AgentSpeak$^+$ agent

related to any of the agent's epistemic states). As needed, during plan execution, the agent can furthermore rely on the POMDP to compute the optimal next step through the use of a probabilistic planning action. Whenever the selected plan (i.e. the one that has been committed as an intention) contains a probabilistic planning action, the corresponding POMDP will be called instead of (directly) relying on the plan library.

Proposition 1 (Proper extension) *An AgentSpeak$^+$ agent $\mathbb{A}^+$ is a proper extension of a classical AgentSpeak agent $\mathbb{A}$.*

Proof An AgentSpeak$^+$ agent $\mathbb{A}^+$ extends a classical AgentSpeak agent $\mathbb{A}$ in three aspects. Firstly, an AgentSpeak$^+$ belief base BB^+ extends an AgentSpeak belief base BB by associating a probability value with each belief atom. Secondly, an AgentSpeak$^+$ agent includes a set of epistemic states EpS. Finally, an AgentSpeak$^+$ plan library PLib^+ extends an AgentSpeak plan library PLib by allowing probabilistic planning plans. If all belief atoms in BB^+ have a probability value of 1, if EpS is empty and if PLib^+ has no probabilistic planning plans, then the AgentSpeak$^+$ agent $\mathbb{A}^+$ reduces to a classical AgentSpeak agent.   □

Proposition 2 (Termination) *Let PLib^+ be a non-recursive AgentSpeak$^+$ plan library and e an event. If there is a relevant plan for e in PLib^+ then either all steps in the plan body will be executed or the plan will fail in a finite number of steps.*

Proof If no applicable plan for e exists in PLib^+, then the execution fails immediately. Otherwise, an applicable plan is selected to deal with e and its plan body is executed. If the applicable plan for e in PLib^+ is a classical AgentSpeak plan then the plan body is a finite sequence of (non-recursive) actions/goals. When an action is encountered it is executed immediately. If a goal is encountered, it is either a test goal, which can be executed immediately by querying the belief base of the agent, or it is an achievement goal. If it is an achievement goal, the same line of reasoning we used so far applies for classical AgentSpeak plans. In addition, because we have that PLib^+ is a non-recursive plan library, we know that after a finite number of subgoals (since the plan library is finite) we will have a subgoal for which the plan only contains actions and/or test goals.

If the applicable plan, or any plans of its subgoals is a probabilistic planning plan then the plan body may also include probabilistic planning actions. By definition, a POMDP used by any probabilistic planning action will always return an optimal policy representing a single action/goal to execute. As before, the resulting action or goal will either succeed or fail in a finite number of steps. $\square$

4   Scenario Discussion

We now show how the scenario from the introduction can be expressed using the AgentSpeak$^+$ framework. Since the material collection scenario relies heavily on optimal planning, the scenario serves as a good example to illustrate the benefits offered by AgentSpeak$^+$ over classical AgentSpeak. We stress though that the purpose of this discussion is not to present an actual implementation, but rather to motivate the merits of the proposed framework to warrant future work on a fully implemented system. As a basis for this future work, we briefly discuss a prototype system at the end of this section which we designed to verify the feasibility of our approach.

4.1   Case Study

We recall that the goal of our truck agent $\mathbb{A}_t$ is to collect materials from factories A, B and C. Then, as discussed in Sect. 3.2, we consider a single epistemic state Φ where its POMDP $\mathcal{M}$ has an action set with 9 subgoals. Each state in the POMDP $\mathcal{M}$ contains information about whether a factory has available materials (denoted as FA, FB and FC), whether our agent has already collected materials from a factory (denoted as MA, MB and MC) as well as the current location of the agent (denoted as LocA, LocB and LocC). For example, the state $s = \{MA, LocA, FA, FB\}$ indicates that the agent has collected material from factory A, is currently at factory A and that material is still available from factories A and B. We define $\mathcal{M}$ in a graphical way to further decompose the state space (i.e. whether material is available at one factory is independent from whether material is available at another). A Dynamic Influence Diagram [1] $\mathcal{D}$ is obtained for efficient computation (shown in Fig. 4). In particular, we decompose the entire state space into a set of chance nodes representing FA, FB, FC, MA, MB and MC while LocA, LocB and LocC are represented by a single chance node Loc. Correspondingly, belief atoms in our agent $\mathbb{A}_t$ describe

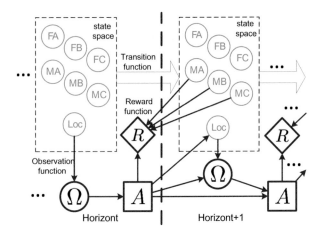

Fig. 4 Graphical representation of POMDP $\mathcal{M}_g$. Here, the entire state space is decomposed into a set of chance nodes. Some causality links for the transition function are omitted for simplicity

the availability of materials (denoted **has(X)**), where materials have been loaded from previously (denoted **loaded(X)**) and the current location of the agent (denoted **at(X)**) for factories $X \in \{a, b, c\}$. As discussed at the end of Sect. 3.1, a revision of the epistemic state is triggered when a belief atom is added/removed and possible additions/deletions of belief atoms are triggered when the epistemic state is revised. The plan library for our AgentSpeak$^+$ agent $\mathbb{A}_t$ contains many plans, including:

```
(P3)    +!start : true ← !callFactories, !collectMaterial.
(P4)    +!callFactories : true ← !check(a), !check(b),
                                   !check(c).
(P5a)   +check(X) : not loaded(X) ← call(X).
(P5b)   +check(X) : loaded(X).
(P6a)   +!waitS1toB : not s1green ← senseSignal; !waitS1toB.
(P6b)   +!waitS1toB : s1green.
(P7)    +!load(X) : at(X) ← pay(X); getMaterial(X),
                             !callFactories.
```

We can describe this agent in more detail. When the agent is initialised, the **start** event is generated. The plan **(P3)** reacts to this event using plan **(P4)**, along with plan **(P5a)** or **(P5b)**, by calling each factory to check if material is available (i.e. it only calls those factories from which material has not been previously loaded). The agent then proceeds with plan **(P3)** by attempting to collect the available materials. The material collection procedure itself was previously described by the probabilistic planning plan **(P1)** from Example 4, which involves a probabilistic planning action using a POMDP to find the optimal plan to execute. A predefined executable plan for moving from the origin location to factory A was described by plan **(P2)** from the same example. With an equivalent plan for moving to factory B, the plan body would consist of moving the agent to the signal, generating a subgoal relevant to plans **(P6a)** and **(P6b)** (i.e. to determine when it is safe to pass the signal), and then proceeding to factory B. Once the agent has completed this action it will check whether it has successfully reached factory B. When the location is sensed, and the agent is at the desired factory, it proceeds by executing plan (P7) to load the required material. First the factory is paid, then the material is loaded and, finally, plan (P4) is again executed to call the remaining factories from which material has not been collected. Importantly, the effect of these subgoals (such as moving, loading and calling factories) modify the agent's beliefs and thus also modify the relevant epistemic state. These new beliefs (and uncertainties) are then taken into account by plan (P1) in order to determine the optimal subgoal for collecting the remaining materials.

In addition to a POMDP $\mathcal{M}_g$, the initial belief state b_1 associated with $\mathcal{M}_g$ is also part of an epistemic state. In the preparation stage (relevant to plan (P4)), $\mathbb{A}_t$ defines correlated belief atoms for each variable in the POMDP's state space, such as FA and FB, and these correlated belief atoms are added to the agent's belief base BB. For each variable, such as FA, the corresponding belief state $b(\text{FA})$ can be defined. When this is completed for all variables, a joint belief state $b(S)$ is derived. For example, the action query(FA) estimates the material availability at factory A and belief atoms, such as available(FA, 0.8) and available(FA, 0.2), express our certainty that the material is available at factory A. The beliefs about other factories, the initially loaded material and the starting location of the agent can be obtained in a similar manner. The belief state b_1 itself can be extracted from the relevant belief atoms according to Definition 11.

This example highlights a number of key benefits offered by the AgentSpeak$^+$ framework. Compared to classical AgentSpeak, we are able to deal with uncertain information. Furthermore, our AgentSpeak$^+$ agent is not fully specified in the design phase; it resorts to probabilistic planning to deal with crucial parts of its execution (e.g. determining the order in which to visit the factories). Compared to a pure POMDP implementation, the AgentSpeak$^+$ framework considerably reduces the overall complexity by relying on domain knowledge encoded on the level of a BDI agent. As such, irrelevant actions such as determining which factories to call and how long to wait at a signal, are omitted from the POMDP dealing with the availability of materials. Furthermore, since planning only happens on-demand, the agent can rely on the simpler plan selection process to ensure maximum reactiveness for most of the subgoals (e.g. when agents also need to achieve goals where uncertainty is absent and/or optimality is not required).

4.2 Implementation Considerations

A prototype implementation of this framework has been developed[4] that extends Jason [5], an open-source implementation of AgentSpeak, with Hugin [1], a proprietary tool for constructing and evaluating Bayesian networks. Equivalent open source

---

[4]By the author Yingke Chen.

alternatives to Hugin include SMILE [10] and its Java API jSMILE. In addition to this, the implementation uses the flexibility of the Jason system to develop parts of the system in Java. The epistemic states, their revision, extraction and derivation have all been defined on the level of Java. As such, the actual agent description in Jason can be mostly agnostic as to the underlying complexity; beliefs are automatically revised and converted into Boolean beliefs as needed, and the only exposure the agent has to the underlying system is through the **ProbPlan** concept. Whenever such a **ProbPlan** action is called, the required POMDP is called through the Hugin API by the Jason interpreter and returns the recommended optimal decision(s). While this prototype proved promising, it suffered from its overall complexity. For example, keeping the beliefs consistent across all three systems is challenging and time-consuming to develop. Still, these tools find optimal solutions for POMDP which can be very time-consuming for even small problems. As a result, the prototype was often considerably slower than plain AgentSpeak. The recent emergency of very capable anytime planning algorithms for POMDP (e.g. [23]) is promising and would be the tools of choice for future implementations. Indeed, by using anytime algorithms an agent could further balance between having reactive behaviour, having quick deliberative behaviour or exhibiting behaviour where the agent can wait if it need not act quickly until an optimal solution is found. Such an algorithm could also be integrated in the Java framework, avoiding the need for expensive API calls to an external tool. Finally, we note that a full evaluation of any implementation would require a problem setting considerably larger than the material collection scenario. Indeed, our framework is developed in such a way that planning happens on demand. In realistic scenarios, however, a large part of the environment can be explored without the need for (near-)optimal actions, i.e. we can rely on simple plan selection rather than planning based on POMDPs. For these reasons, the development of a full implementation, as well as its thorough evaluation, is left for future work.

5   Related Work

There have been several approaches to modelling uncertainty in multi-agent systems. In [7] a graded BDI approach is proposed that uses uncertain beliefs (as probabilities) and graded preferences (as expected utilities) to rank plans. However, the theoretical nature of this work has so far precluded the development of any practical implementations. In this work, we instead propose an extension based on a widely used agent-oriented programming language. Our work is based on epistemic states, similar to [8], where the concept of an epistemic state is introduced to model uncertain perceptions. Similar to their work, Boolean belief atoms (e.g. propositional statements) can be derived from epistemic states to support further reasoning. Still, the work in [8] only focuses on MDPs, i.e. it cannot be used in the more natural setting of partially observable environments. A similar approach has been used in [2] where the authors model different forms of uncertainty as distinct epistemic states. This allows a single agent to reason about different forms of uncertainty in a uniform way. However, the work's main focus is on the representation of the beliefs and their commensurability and does not provide first-principles planning under uncertainty, nor do they exploit any of the facets of the decision theoretical model (e.g. rewards or penalties).

Autonomous agents have to make rational decisions to pursue their goals (i.e. selecting appropriate plans) in a stochastic environment. Markov Decision Processes (MDP) and Partially Observable MDPs (POMDPs), are popular frameworks to model an agent's decision making processes in stochastic environments. In [22] a theoretical comparison of POMDPs and the BDI architecture identifies a correspondence between desires and intentions, and rewards and policies. The performance and scalability of (PO)MDPs and the BDI architecture are compared in [24]; while (PO)MDPs exhibit better performance when the domain size is small, they do not scale well since the state space grows exponentially. The BDI architecture (which uses a pre-defined plan library) has better scalability at the cost of optimality, making it applicable to significantly larger state spaces. Nevertheless, future research showed that BDI and MDP are closely linked. Indeed, in [25] the relationship between the policies of MDPs and the intentions in the BDI architecture is discussed. In particular, it shows that intentions in the BDI architecture can be mapped to policies in MDPs. This in turn led to some work in the literature on hybrid BDI-POMDP approaches. In [18] an algorithm was proposed to build AgentSpeak plans from optimal POMDP policies. However, most characteristics of the original BDI framework are not retained in such hybrid approaches. In contrast, our approach embeds POMDPs in the traditional BDI agent framework. Normal BDI execution is used by default, with the POMDP component allowing an agent to generate new plans on-demand during execution. Extending the BDI architecture with more elaborate planning techniques has also been investigated in the literature. In [21], the authors present a formal framework to integrate lookahead planning into BDI, called CANPLAN. Similarly, in [16], the authors integrate classical planning problems into the BDI interpreter to allow an agent to respond to unforeseen scenarios. However, neither approach considers issues related to uncertainty.

6   Conclusions

In this work we proposed the AgentSpeak$^+$ in which we extend the belief base of an agent by allowing it to be represented by one or more epistemic states. An essential part of each epistemic state is a POMDP, which allows us to model the domain knowledge of the partially observable environment and from which we can compute optimal actions when needed. In addition, this allows us to deal in a straightforward way with uncertain information in the environment. To keep the computational complexity low, AgentSpeak$^+$ extends AgentSpeak, an agent-programming language based on the BDI paradigm where planning is reduced to the simple task of plan selection. By adding actions to perform on-demand planning, the resulting AgentSpeak$^+$ can both offer good responsiveness while at the same time providing the option for near-optimal planning when needed through the POMDP component. For future work, we plan a full evaluation of our approach, both compared to classical BDI implementations and pure POMDP implementations. We furthermore plan an extension where knowledge from other agents (which are only trusted to some degree) can be employed to improve the domain knowledge currently encoded in the POMDP component.

References

1. Andersen, S.K., Olesen, K.G., Jensen, F.V., Jensen, F.: HUGIN: a shell for building bayesian belief universes for expert systems. In: Proceedings of the 11th International Joint Conference on Artificial Intelligence (IJCAI'89), pp. 1080–1085 (1989)
2. Bauters, K., Liu, W., Hong, J., Sierra, C., Godo, L.: Can(plan)+: extending the operational semantics of the BDI architecture to deal with uncertain information. In: Proceedings of the 30th Conference on Uncertainty in Artificial Intelligence (UAI'14), pp. 52–61 (2014)
3. Bellman, R.: A markovian decision process. Indiana Univ. Math. J. **6**, 679–684 (1957)
4. Blum, A.L., Langford, J.C.: Probabilistic planning in the graphplan framework. In: Recent Advances in AI Planning, pp. 319–332. Springer, Berlin (2000)
5. Bordini, R.H., Hübner, J.F., Wooldridge, M.: Programming Multi-agent Systems in AgentSpeak using Jason. Wiley-Interscience (2007)
6. Braubach, L., Lamersdorf, W., Pokahr, A.: JADEX: implementing a BDI-infrastructure for JADE agents. EXP - in search of innovation **3**(3), 76–85 (2003)
7. Casali, A., Godo, L., Sierra, C.: A graded BDI agent model to represent and reason about preferences. Artif. Intell. **175**(7–8), 1468–1478 (2011)
8. Chen, Y., Hong, J., Liu, W., Godo, L., Sierra, C., Loughlin, M.: Incorporating PGMs into a BDI architecture. In: Proceedings of the 16th International Conference on Principles and Practice of Multi-Agent Systems (PRIMA'13), pp. 54–69 (2013)
9. de Silva, L., Sardiña, S., Padgham, L.: First principles planning in BDI systems. In: Proceedings of the 8th International Joint Conference on Autonomous Agents and Multiagent Systems (AAMAS'09), pp. 1105–1112 (2009)
10. Druzdzel, M.J.: SMILE: structural modeling, inference, and learning engine and GeNIe: a development environment for graphical decision-theoretic models. In: Proceedings of the 16th National Conference on Artificial Intelligence (AAAI'99), pp. 902–903 (1999)
11. Hansen, E.A.: Solving POMDPs by searching in policy space. In: Proceedings of the 24th Conference in Uncertainty in Artificial Intelligence (UAI'98), pp. 211–219 (1998)
12. Jennings, N.R., Bussmann, S.: Agent-based control systems. IEEE Control Syst. Mag. **23**, 61–74 (2003)
13. Kaelbling, L.P., Littman, M.L., Cassandra, A.R.: Planning and acting in partially observable stochastic domains. Artif. Intell. **101**(1), 99–134 (1998)
14. Ma, J., Liu, W.: A framework for managing uncertain inputs: an axiomization of rewarding. Int. J. Approx. Reason. (IJAR) **52**(7), 917–934 (2011)
15. McArthur, S.D., Davidson, E.M., Catterson, V.M., Dimeas, A.L., Hatziargyriou, N.D., Ponci, F., Funabashi, T.: Multi-agent systems for power engineering applications - Part I: concepts, approaches, and technical challenges. IEEE Trans. Power Syst. **22**(4), 1743–1752 (2007)
16. Meneguzzi, F., Luck, M.: Declarative planning in procedural agent architectures. Exp. Syst. Appl. **40**(16), 6508–6520 (2013)
17. Meneguzzi, F., Tang, Y., Sycara, K., Parsons, S.: On representing planning domains under uncertainty. In: Proceedings of the 3rd Information Theory and Applications Workshop (ITA'10) (2010)

18. Pereira, D., Gonçalves, L., Dimuro, G., Costa, A.: Constructing BDI plans from optimal POMDP policies, with an application to agentspeak programming. In: Proceedings of Conferencia Latinoamerica de Informtica (CLI'08), pp. 240–249 (2008)
19. Rao, A.S.: Agentspeak(l): BDI agents speak out in a logical computable language. In: Proceedings of the 7th European Workshop on Modelling Autonomous Agents in a Multi-Agent World (MAAMAW'96), pp. 42–55 (1996)
20. Rao, A.S., Georgeff, M.P.: An abstract architecture for rational agents. In: Proceedings of the 3rd International Conference on Principles of Knowledge Representation and Reasoning (KR'92), pp. 439–449 (1992)
21. Sardina, S., Padgham, L.: A BDI agent programming language with failure handling, declarative goals, and planning. Auton. Agents Multiagent Syst. **23**(1), 18–70 (2011)
22. Schut, M., Wooldridge, M., Parsons, S.: On partially observable MDPs and BDI models. In: Proceedings of the UK Workshop on Foundations and Applications of Multi-Agent Systems (UKMAS'02), pp. 243–260 (2002)
23. Silver, D., Veness, J.: Monte-carlo planning in large POMDPs. In: Proceedings of the 24th Annual Conference on Neural Information Processing Systems (NIPS'10), pp. 2164–2172 (2010)
24. Simari, G.I., Parsons, S.D.: On approximating the best decision for an autonomous agent. In: Proceedings of the 6th Workshop on Game Theoretic and Decision Theoretic Agents (GTDT'04), pp. 91–100 (2004)
25. Simari, G.I., Parsons, S.: On the relationship between MDPs and the BDI architecture. In: Proceedings of the 5th International Joint Conference on Autonomous Agents and Multiagent Systems (AAMAS'06), pp. 1041–1048 (2006)

Ant-Based System Analysis on the Traveling Salesman Problem Under Real-World Settings

Gloria Cerasela Crişan, Elena Nechita and Vasile Palade

Abstract Many optimization problems have huge solution spaces, deep restrictions, highly correlated parameters, and operate with uncertain or inconsistent data. Such problems sometimes elude the usual solving methods we are familiar with, forcing us to continuously improve these methods or to even completely reconsider the solving methodologies. When decision makers need faster and better results to more difficult problems, the quality of a decision support system is crucial. To estimate the quality of a decision support system when approaching difficult problems is not easy, but is very important when designing and conducting vital industrial processes or logistic operations. This paper studies the resilience of a solving method, initially designed for the static and deterministic TSP (Traveling Salesman Problem) variant, when applied to an uncertain and dynamic TSP version. This investigation shows how a supplementary level of complexity can be successfully handled. The traditional ant-based system under investigation is infused with a technique which allows the evaluation of its performances when uncertain input data are present. Like the real ant colonies do, the system rapidly adapts to repeated environmental changes. A comparison with the performance of another heuristic optimization method is also done.

Keywords Combinatorial optimization · Uncertainty modeling · Traveling Salesman Problem · Ant Colony Optimization

G.C. Crişan (✉) · E. Nechita
Vasile Alecsandri University, Bacău, Romania
e-mail: ceraselacrisan@ub.ro

E. Nechita
e-mail: enechita@ub.ro

V. Palade
Coventry University, Coventry, UK
e-mail: vasile.palade@coventry.ac.uk

© Springer International Publishing Switzerland 2016
I. Hatzilygeroudis et al. (eds.), *Combinations of Intelligent Methods and Applications*, Smart Innovation, Systems and Technologies 46,
DOI 10.1007/978-3-319-26860-6_3

39

1 Introduction

Decision under uncertainty is a research domain with documented traces within the second half of the last century. Stochastic programming [1] and Optimization under probabilistic constraints [2] are the first of such research directions stemming from the same assumption: the distributions of the random variables are exactly known.

Later on, the uncertainty was modeled in a broader way, as the data inexactness was defined and treated outside the Probability Theory. During 1965–1975, the inexactness was approached with different fundamental ideas that developed important optimization domains. One direction was to model "… *problems in which the source of imprecision is the absence of sharply defined criteria of class membership rather than the presence of random variables.*" [3], by the well-known Fuzzy Sets framework. If no assumption is made on inexact data, except for their bounds, then a problem can be approached using a Robust Optimization methodology. Pioneering works considered the *generalized* linear programming problem [4], or the *proximate* linear programming problem [5], for the situations when "…*there exists a real need of a technique capable of detecting cases when data uncertainty can heavily affect the quality of the nominal solution, and in these cases to generate a* reliable *solution, one which is immuned against uncertainty.*" [6]. During the last years of the past century, two independent teams grounded the Robust Optimization methodology [7, 8]—branch of Optimization with broad applications and deep mathematical foundations. A modern approach, inspired by the broad developments of large data collections which can be inconsistent, is presented in [9]. The authors assess the risk of using erroneous data by the solving methods designed for correct data.

The current problems exhibit compound complexity and challenge us from multiple points of view. Firstly, their *high dimensional and/or high volume search spaces* make impractical the attempt to exactly solve them. The only affordable approach is to accept "good enough" solutions by using approximate or heuristic methods. Secondly, the intricate complexity of the problems (having *many, deep and partially known connections between variables*) makes the computing time increase to the level of intractability. By considering the decomposition into several simplified problems and reconstructing the initial solution from the solutions to the simpler problems (be they optima), rarely the final solution is optimal [10]. Other feature of the complexity is the data uncertainty: as the problem is rich in parameters, variables, restrictions and equations, some degree of data uncertainty is expected to produce dissimilarity between the model and the real problem. Thirdly, the real problems never are static and isolated: when seen as a process, they permanently interact with other neighboring processes (neighbor is used here for describing other concurrent and connected processes). This last feature expresses the *dynamicity*, adding up new difficult aspects to an already complex model.

This paper investigates how a heuristic method designed for a specific problem behaves when applied to a more complex problem variant that displays two supplementary layers of difficulty. The application under discussion is an ant-based system, which was adapted to handle two new problem characteristics: uncertainty and dynamicity. For testing it, we used 24 instances with thousands of vertices of the Traveling Salesman Problem (TSP, a well-known Combinatorial Optimization Problem—COP). The results are compared with those obtained when the original software package solved the same 24 instances. Our work shows that the ant-based implementation can easily handle the dynamic uncertainty, being able to deliver results having the same quality as in the classical case. Moreover, the instances are approached with another classical, heuristic-based method in order to compare their stability.

When facing specific situations, decision-makers would like to rely on sound, rapid intelligent/expert systems. This means that the solving methods have to provide good solutions when tackling high volumes of uncertain, dynamic and complex data that have many intricate restrictions. For example, if a manager is aware of the scalability, resilience and fault-tolerance of several support systems on the market, she/he can take an informed decision when the company decides to buy such products.

The quality of the support system has an even greater importance when the organization using it is large, or works in sensitive areas. For example, the sea transportation deals with weather uncertainty, environmental risks in case of accidents and implies moving outsized expensive goods over large paths. Due to these characteristics, ship routing decision systems must intelligently assimilate many specific regulations and integrate detailed systems, like GIS, into academic models [11].

Project managers from the software industry are interested in high-quality products, with efficient behavior when complex situations appear. When designing assisting applications for ocean liners, for example, to improve the ship performance while minimizing the risk of structural damage can be the major strategic objective. At operational level, the shipmaster can be mainly interested in predicting the ship motions and structural loads due both to immediate weather conditions and to possible changes in course and speed [12].

Consequently, this paper would be useful for both customers and designers of software applications dedicated to solving difficult COPs: as the heuristic method we have exercised has a stable and reliable behavior. Efficient implementation of this method might lead to software modules which could be successfully integrated in decision-support systems.

This work is structured as follows: the next section describes our problem; Sect. 3 introduces the method we have used; in Sect. 4 we detail the method; Sect. 5 discusses our results, and we draw the conclusions in Sect. 6. The Annex presents the detailed results of our simulations.

2 The Problem

Our paper uses the experimental analysis in order to asses how an application designed for static, exact problems behaves when dynamicity manifests in an uncertain way. As we intended to focus on an already stable and fast application, we firstly tested from the stability point of view two heuristic implementations: the chained Lin-Kernighan [13] from [14], and the ACOTSP [15], both designed for the academic Traveling Salesman Problem (TSP). ACOTSP, the most stable software package was then altered in order to allow repeated environmental changes, modeled as random modifications of the data.

2.1 The Traveling Salesman Problem

TSP is one of the most important problems in Operational Research [16, 17]. It has documented historical traces back to Leonhard Euler; it was formalized in 1930 by Karl Menger and became popular among mathematicians in the 1940s [18]. Since then, the TSP was largely and intensively investigated, as it is easy to formalize and to represent, and it models many current real-life problems [19].

When solving the TSP, one has a map with n places and knows their pairwise distances, and is asked to find one shortest cyclic tour through all the places. More formally, the TSP seeks for a least cost Hamiltonian cycle in a weighted graph. From the computational complexity point of view, the TSP is an $\mathcal{NP}$-hard problem, so a polynomial time exact algorithm for the worst case is unlikely to be found [20].

TSP applications can be found in quite diverse areas of the human activity. Some of them deal with large or very large instances. This is the reason why researchers are interested in developing TSP algorithms to provide good solutions for this type of instances. A theoretical study on some methods and refinements suitable to approach large instances is given in [21], while particular topics can be easily found in literature.

For example, there are practical problems, such as those involving circuit-board drilling, board cutting or chip manufacture [22] which can be modeled by TSPs with millions of nodes. Implementations of the Chained Lin-Kernighan heuristic [13] have been used in this direction. In the field of genomics, where the nature of the working problem provides large scale combinatorial optimization problems, TSP algorithms are used to generate maps in genome sequencing [23, 24].

Conversely, tools of molecular biology could be used to provide solutions for TSP, with graphs encoded in molecules of DNA and computations performed with standard protocols and enzymes. A pioneering view on this matter is given in [25]. Although DNA molecules could store immense amounts of information and massive parallel searches could be imagined, researchers still have to study if and how this potential can be exploited [26].

For large TSP instances, the researchers focused on finding good relaxation methods for effectively decreasing the solution space. Held and Karp proposed the most known relaxation algorithm [27, 28], with very tight bounds in practice and, therefore, included in the Concorde solver [29] (which currently is the best TSP exact solver). Besides the Concorde solver, other software products are available for research use [30, 31]. The approximate methods for TSP received intense attention, as they provide results with an a priori known quality [32]. There also are heuristic and meta-heuristic approaches dedicated to TSP, showing a broad interest from researchers and a wide suitability for investigations: local search heuristics [33, 34], Simulated Annealing [35], and several nature-inspired methods. Among the heuristic methods, the 3-opt and the Lin-Kernigan methods [33] and Marco Dorigo's Ant Colony Optimization [36] particularly interest us in this paper.

For testing, comparing and developing new applications, several standard collections of TSP instances are available [37, 38]. For this study, we have chosen, from [38], the 24 TSP symmetric Euclidean 2D instances with a number of vertices between 1,000 and 6,000. Their names are listed in the second column of Table 1. Other available instances integrate modern features like GIS services, able to connect research from Transportation, Logistics, Geoinformatics, etc. [39].

Table 1 LK versus ACOTSP stability

#	Instance	100*CV_LK	100*CV_ACOTSP
1	d1291	0.4490	0.0405
2	d1655	0.2744	0.1334
3	d2103	0.3948	0.0388
4	fl1400	0.0979	0.2687
5	fl1577	1.5141	0.0672
6	fl3795	0.4205	0.5516
7	fnl4461	0.0313	0.0730
8	nrw1379	0.0449	0.1024
9	pcb1173	0.1393	0.1120
10	pcb3038	0.0688	0.1047
11	pr1002	0.1565	0.1695
12	pr2392	0.0658	0.3282
13	rl1304	0.2776	0.1023
14	rl1323	0.1146	0.0808
15	rl1889	0.1499	0.1730
16	rl5915	0.1594	0.1573
17	rl5934	0.1205	0.2173
18	u1060	0.0932	0.1268
19	u1432	0.1062	0.0726
20	u1817	0.1764	0.1579
21	u2152	0.1260	0.1754
22	u2319	0.0250	0.0450
23	vm1084	0.1009	0.0488
24	vm1748	0.0321	0.1627
	average	**0.2141**	**0.1462**

2.2   TSP Under Uncertainty

It is obvious that the recent advances in engineering and communication technologies have determined innovative ways of developing human activities. In turn, these have generated contexts which are more and more complex. Therefore, the models for these contexts require appropriate mathematical and computational tools.

A relevant example on this line is the field of transportation. Today, quick and easy transportation is seen as an essential part of modern society. Users can order mobility services via Internet, by using smart devices. The positioning systems are used to support a huge and dynamic network of vehicles. We face growing urbanization and traffic volumes, which entails environmental pollution, increasing fuel consumption, accidents and other expensive social costs. Moreover, there are a lot of uncertain factors to be considered, for example in travelers' presence, in travel time, in merchandise demand. In order to describe such systems with a reasonable degree of accuracy, the corresponding models require appropriate handling of uncertain information. As researchers stress [40], the presence of uncertainty arises two important issues: to model it, for a proper representation of the real-world situations, and to develop computationally tractable applications for the models under discussion.

As there are different types of formalizing uncertainty, a multitude of approaches have been proposed and investigated by researchers. The classical models for routing optimization problems (TSP, Shortest Path Problem and Vehicle Routing Problem) have already known numerous extensions and variations. The aim of this section is to give an overview over the most substantiated ones and to underline the efficiency of heuristics versus other methods.

The Probabilistic Traveling Salesman Problem (PTSP), introduced by Patrick Jaillet in [41] is a variant of TSP in which only a subset of the nodes may be present in any given instance of the problem. In other words, in PTSP the probability of each node being visited is between 0 and 1. The goal, when approaching PTSP, is to find an a priori tour of minimal expected length, with the strategy of visiting the present nodes in a particular instance in the same order as they appear in the a priori tour. The literature presents several approaches of PTSP, including exact branch-and-bound [42] and branch-and-cut [43] algorithms, for instances with at most 50 vertices. But most researchers focus on heuristics such as nearest-neighbor [44], 2-opt and *1*-shift [45] and on metaheuristics, among which ACO [46] and Honey Bees Mating Optimization (HBMO, [47]) proved to be very efficient. A recent survey and considerations on the PTSP may be found in [48]. In [49], a closer look and review on PTSP are done from the perspective of approaching it with ACO and local search. Other uncertain models for the TSP are the Robust TSP [50], or the Fuzzy TSP [51].

Another direction to extend TSP comes from the necessity to model dynamic or adaptive features. For example, if a driver is receiving disadvantageous, real-time information about the road to a city where he is supposed to move, he may decide to shift towards another city. In terms of modeling, this scenario corresponds to a system in which the cost of the decision in not known a priori but assigned when the decision must be taken [52]. Such dynamic models could be approached with Dynamic Programming (DP) tools [53], but numerous studies have shown their limited applicability [19]. The large dimensionality of the DP implementations could be overcome with Approximate Linear Programming [54]. Various models of TSP with uncertain arc costs are proposed in [55–57]. A new perspective on uncertain optimization problems is presented in [58]: the goal of the research is to use a measure of similarity on previous instances to predict the quality of the solutions to the following "close" instances. This idea can be founded in this paper, too.

Another direction that involves uncertainty in TSP refers to the versions handling online data. Online optimization may be needed, for example, in situations like rescue operations during natural disasters, when the objectives (cities, buildings, etc.) to be visited are not known a priori and are revealed on the way. A general framework for the online versions of the TSP and optimal 2-competitive polynomial time online algorithms, based on re-optimization approaches, for TSP with Release Dates and TSP with Flexible Service are presented in [59]. An extensive literature review over the online TSP is further presented in [60]. A dedicated research project—Classical and Online TSP—has been launched in 2011 at MIT [61]. The goal of the project was to design and analyze rigorous algorithmic solution strategies for online version of classical combinatorial optimization problems (TSP, Hamiltonian path) with incomplete and uncertain input streams on one hand, and with time-sensitive objectives on the other hand.

Looking from the above perspectives, the experiment presented in Sect. 4 falls into the area of TSP versions with uncertain arc-related information, as the implemented ACO algorithm works on randomly modified pheromone matrices. This change in the ACO algorithm thus simulates uncertainty in arcs, while the vertices-related information suffers no alteration.

2.3 Preliminary Tests

The Lin-Kernighan implementation from [14] was called with a variable random seed, the ACOTSP application was downloaded from [15] and was locally used. Both methods ran 10 times on each instance, and their stability was individually assessed by the averaged coefficient of variation of the 10 solutions they provided (presented in Table 1).

The Lin-Kernigan optimization procedure (that falls into the class of local search algorithms) involves swapping sets of k edges to generate feasible, possibly better solutions. It generalizes the 2-opt and 3-opt basic moves by dynamically deciding the value of the parameter k and seeking for better k-opt moves. The method proved to be adaptive and highly effective. However, its usefulness depends on the graph that models the problem—for example, it is less efficient for the instances that are sparse or heavily clustered.

The ACOTSP package implements Ant System, Elitist Ant System, $\mathcal{MAX}$-$\mathcal{MIN}$ Ant System, Rank-based version of Ant System, Best-Worst Ant System, and Ant Colony System for the symmetric TSP [15]. The goal of this package is to unify the most important ACO algorithms under ANSI C language for Linux, offering the research community a reasonably high performing access under a GPL license.

At the beginning of the algorithm, the artificial ants are randomly deployed in the vertices of the graph modeling the problem. The way an artificial ant chooses the next move when constructing its path is a balance between exploration and exploitation: the closer vertices and also the edges with intense pheromone are preferred, but the algorithm is not deterministic. All the ants construct their path in parallel, and deposit the pheromone according with the problem's goal and restrictions. The process is repeated until the stopping conditions are met. At the end, the best path (from the start) is the solution to the problem.

Table 1 presents the percent coefficients of variation (%CV) for the ten best solutions provided for each instance by both solving methods. The values in the last two columns of Table 1 (and also the corresponding average values over all instances, presented in the last line) show that ACOTSP could perform better related to our goal of introducing uncertainty. Because we intend to study a resilient and stable implementation, we turned our attention towards ACOTSP.

One of the current challenges when exploiting complex software applications is how to predict and control their behavior when some changes (whose outcome is known when they perform alone) manifest together. This is the wider problem that we address in this paper. The way we enrich the TSP and the experiment we have performed in order to study its behavior are presented in the next section.

3   The Method

Co-evolution and adaptation are two effects of living organisms' evolution. Genetics integrated with the theory of evolution explain the diversity and the complexity of life as we know it (and we continually discover) on our planet. Humans are just one species on Earth, with an extremely short history. Turning our attention on how other species solved their problems could inspire us in solving ours. Such a case is presented in this section.

A colony of ants is a successful *eusocial* organization of life: ants account for 15–25 % of the terrestrial animal biomass, colonizing almost the entire Earth [62]. Eusociality is the highest level of social behavior; it includes but is not limited to: cooperative brood care, overlapping generations within a colony of adults, and a division of labor into reproductive and non-reproductive groups [63]. It is amazing how so small and almost-blind insects manage to live and to evolve in so various environments. The ant colony, as many other living systems, is in continuous and intimate connection with the environment. The ants modify the environment they live in, and the environmental changes force the colony to adapt.

The ant colony behaves like a super-organism, manifesting several emergent properties that miss at individual level: the entire colony survives when attacked, discovers the shortest path from food to nest, and is able to move on and to fight for new territories, even if these actions could not be independently completed by individual ants. Moreover, the emergence is achieved without central coordination, only on the basis of local interactions between individuals. The entire system displays a self-organizing structure: some individuals could fail in reaching their goals or they even die, but the colony successfully develops.

Ant Colony Optimization (ACO) is a constructive meta-heuristic for solving COPs, inspired by the way a colony of ants succeeds in finding the shortest path between two points on the ground [64]. The effectiveness of the (natural) colony is a result of concurrent local interactions: each ant can be seen as an agent that perceives the environment, decides on the next move according to its goal and to the current local situation, and consequently acts. An ant indirectly communicates with its peers by laying a chemical substance (called pheromone) on the ground, during its walk. The ants that later walk through are able to perceive the deposited pheromone and more likely follow the chemical trace.

Like the real ants, the artificial ants form a colony of cooperating agents that solve a specific COP. The first problem solved with an ACO implementation was the TSP [36]. A collection of artificial ants repeatedly construct complete tours of the TSP instance. At any iteration, the ants construct in parallel their complete tour. After a tour is completed and its length is known, all its edges receive a quantity of pheromone directly proportional with the tour quality. If an edge is included in many very good tours, it receives high quantity of pheromone and becomes attractive for the following ants. The algorithm is repeated until the exit-condition is met, and the output is the shortest tour founded since the start. Depending on the problem, the ACO implementation could use supplementary modules, designed for obtaining better solutions: local search, pheromone enforcing on specific paths, etc. [65].

The $\mathcal{MAX}$-$\mathcal{MIN}$ Ant System (MMAS) implementation chosen here has very good results for symmetric TSP. It is elitist in the sense that only one best ant is allowed to deposit pheromone, it is adaptive as it avoids search stagnation and it is dynamic as it favors exploration in its early stages. By calling the 3-opt local search procedure for each ant, after each tour completion, the MMAS is a hybrid method, as it integrates both constructive and local-search heuristics.

The adaptive biological models inspired humans in solving their problems. Some of the self-healing materials from Engineering, self-adaptive traffic control systems from Transportation and Logistics, targeted drug delivery based on nanosensors in Medicine, or ad-hoc networks and multi-agent systems from Computer Science are based on the same ideas of co-evolution and adaptation.

Our investigation follows the same path: we here apply biologically-inspired methods to solve complex problems and study some of their implementations' characteristics. Firstly, we ran the original MMAS on the selected instances. Secondly, we modified the application and introduced four new parameters, for setting the frequency of the alterations, their extent and the level of uncertainty. In all details, our experiment is described below.

4 Experimental Design and Implementation

As we intended to address difficult TSP instances, we selected from [38] the symmetric Euclidean 2D instances having more than 1,000 vertices. These instances are presented in the grey cells of the first column from Table 5. These instances were approached using the ACOTSP application from [15], but also with our modified version of it, able to tackle uncertainty and dynamicity.

Our previous investigations showed that small modifications in the classic instances result in small modifications in the results [66–68]. As we intend to study complex, inter-related phenomena, the uncertainty and the dynamicity are here collectively addressed. The model is the human society, where social trends could influence the global evolution. This idea is consistent with other preoccupations from Biology (epidemic models) or Sociology (ad-hoc street protests). We decided to maintain the instances and to change the software package, by repeatedly introducing modifications in the data structures containing the agents' view, namely the matrix containing the pheromone loads on edges. The artificial ant colony is also enriched with a technique which allows the user to evaluate the system's performance while it tries to adapt to environmental modifications.

Using the variable `iteration` from the original MMAS as timer, we introduce the new parameter `epoch` that sets the number of iterations when the system uses static pheromone matrix. When the next epoch starts, all the agents "change their opinion", as the values from this matrix non-deterministically change. This *dynamicity* is maintained until the algorithm ends, and we counted (on average for 10 executions) from 2 (for rl5934, the instance with the largest dimension) to 19 such epochs (for u1462).

After each epoch ends, we produce a "shake" using three new parameters: `p_nod_ch` (the percentage of the vertices chosen to suffer modifications), `p_inten` (the percentage of the edges incident to the chosen vertices that have their pheromone load modified), and `ampl` (the amplitude of the modification, showing the length of the interval where the new pheromone value could be placed). For enriching the complexity, every parameter introduces *a direction of uncertain behavior*: the vertices and the edges are randomly chosen, and the new pheromone values are randomly distributed in the interval defined by `ampl`, centered in the old pheromone value.

The idea of data shaking is not new, as it appears for example in the Variable Neighborhood Search specification and is successfully integrated in other heuristic approaches (chained Lin-Kernighan [13], Tabu Search [69], Simulated Annealing [70, 71], etc.).

Table 2 Parameter values for the MMAS application

#	Name	Value
1	Number of ants	25
2	Number of tries	10
3	Alpha (influence of pheromone trails)	1
4	Beta (influence of heuristic information)	2
5	Rho (evaporation rate)	0.2
6	Number of nearest neighbors	20

These four new parameters model a systematic alteration, applied in the same way to all the instances: the number of "shaken" edges depends on the dimension of the instance; the distance between the new pheromone value and the old one has the same percentage upper limit.

During our computational investigations, we have used epoch=60, p_nod_ch= 4%, p_inten=10%, while ampl took five possible values: 0%, 10%, 20%, 30%, and 40%. These values determine five runs for each instance and correspond to: classic MMAS approach for exact TSP, and four uncertain TSP with [0.9*ov, 1.1*ov], [0.8*ov, 1.2*ov], [0.7*ov, 1.3*ov], and respectively [0.6*ov, 1.4*ov] as interval for the new possible pheromone value, where ov is the old pheromone value deposited on the considered edge. The five values for ampl provide the group of five lines for each instance from Table 5. The first line from the group (in gray) is the result of the initial MMAS. The next four lines from the group are the results of the modified application.

According to the considerations in [72] regarding the evaluation of the stochastic algorithms performances on benchmark problems, we have reported the average values of the performances that we have recorded during the experiment.

The MMAS original parameters were set to their implicit values (presented in Table 2), except for the execution time. As our instances are big, we set 60s running time for the 21 instances having less than 4,000 vertices and 120s running time for the three remaining (in Table 5, they have an asterisk following their names). We executed the application for 120 times (five times for each instance), and the total computation time was 22.5 hours, on a dual-core processor at 2.5 GHz frequency and 2GB RAM memory. During the tests, one core was assigned to the application only. The memory load went up to 80% for the top 3 biggest instances, and for the other instances it was about 50%.

5   Results and Discussion

The original MMAS iterates the same method 10 times and delivers three files with results and additional information on its evolution. From these files we gathered some statistics into Table 5, and we added the optimum value for each instance (column 2), taken from [38], and a new variable RS (Recovery Speed), presented in the third column. We introduced this new metric in order to assess the adaptability of the application, and its capacity to deliver better solutions even when all the agents have an altered perception on the environment. RS counts how many times (on average in 10 executions) after the first change of the pheromone matrix, the system succeeds in finding a better solution [73]. As we wanted to highlight the recovery feature, the cells with bigger RS than the classic MMAS approach have gray background. The remaining columns hold: the best solution among the 10 best solutions from each set of 10 executions (Best), the worst solution among them (Worst), the average value of the 10 best solutions (Av B), the average value of the iteration that found the best solution (Av I), the average moment of time from the beginning of the execution when the best solution was found (Av T), the standard deviation for the 10 best solutions (SD B), and the standard deviation for the 10 iterations that found the best solutions (SD I). In the last seven columns, the least value for each group of five lines is boldfaced. The last five lines from the table are the averaged values for all instances, taken on corresponding groups (a group corresponds to a specific value for the ampl variable).

The RS column globally shows that the artificial Ant System succeeds in maintaining its feature of continuous solution improving: in 53 from the 96 altered cases (more than 55 %) it has at least the same value like the RS of the original application; from the remaining ones, 17 times it has less than a 10 % distance, 14 times has a distance between 10 and 20 %, 10 times is between 20 and 30 % far, and in just two cases (both for the fnl4461 instance) is between 30 and 35 % far from the RS in the classic case. The last five cells from the RS column show that, on average, the small perturbations are well handled and the system quickly and repeatedly straightens and finds better solutions; when the perturbations have larger amplitudes, the system slows in finding better solutions, but the speed is within the same range as the speed for the classic MMAS: the average of the values from the last four cells is 2.185, which is better than the RS for the initial application.

The columns 4 to 6 refer to the quality of the best solutions. The fourth column shows an almost-perfect balance between the five cases considered: for 7/8/9/7/5 instances, the 0/10/20/30/40 modification performs better (we have instances where multiple cases outcome the same values). For the fifth column we have the repartition 8/2/6/1/8, and for the sixth this is 6/2/10/0/6. These two series of results show that reasonable changes do not affect the quality of the best outcome, so the application can be successfully used even in the case when data alterations are expected. When there is no time to wait for multiple executions, the application performs alike, whether or not a sequence of uncertain modifications in data is present. We can state that, within our experiment, the modifications do not alter the quality of the solution;

it is interesting that the mid-amplitude alteration, the same that is lazy in evolution, performs best: it provides 6 best values in column 5 and 10 best values in column 6. It is also interesting that the extreme situations are also very good: the classic MMAS and also the version with large interval of possible pheromone values provide also good results for the columns 5 and 6.

The columns 7 and 8 refer to the speed of the application, showing the iteration and the moment of time when the best solution is found. Globally, these values form a tight interval, and on average they are balanced: 2/6/4/6/6, and 4/2/8/8/2 respectively, with a small bias towards the 30 % amplitude. As the solutions in this case are bad, they are found in early stages and so the algorithm spends the last seconds with no improvements.

The last two columns are dedicated to the standard deviation for the best values and for the iterations that found them. They display information on the supplementary struggle of the algorithm to face the change (the initial application has the best stability), but the most stable from the iteration point of view are—again interestingly—the mid- and high alterations.

In order to identify one or maybe more patterns in the behavior of the application when is altered as described, we performed a correlation analysis. The computation of the correlation coefficients for the pairs of columns: (Av B, Av T) and (Av I, Av T) revealed in all cases no significant linear correlation. For example, an improvement of the tour lengths (values from Av B) is not achieved faster (column Av T). Smaller execution times of the application do not correlate with the moments (measured by the average iteration, Av I) at which better solutions are found. These results are not surprising, given the random nature of the ACO paradigm: the application was expected to behave this way.

Based on the data displayed in Table 5 we could state that the most favorable behavior is exhibited by the application altered with `ampl=20 %` (a20 %). In order to verify this outcome, we have performed 30 new executions for the instance pr2392 only and recorded, as usual, the best solutions (denoted with Best) and the moment of time when Best is found (denoted with Time). For these two random variables, measures of central tendency (mean, median and midrange), of variability (standard deviation, range and interquartile range) and symmetry (skewness) have been computed: for Best in Table 3 and for Time in Table 4. In both tables, the lines with mean values that are smaller than the mean values for the classic MMAS (in bold font) are shaded in grey. As the reader can see, the lines corresponding to a20 % and a40 % are emphasized in both tables, but the line in Table 3 also holds the minimum value in column Min (the smallest "best solution" has also been recorded for a20 %). Looking at the values in columns Stdev, Range or IQR for a20 %, in Table 3 again, we may interpret that these reflect the possibility that a20 % facilitates, namely for an enhanced exploration of the solutions space. Moreover, the slightly (left) skew of the distribution may point the same thing.

Table 3 Statistics for **Best** on 30 executions on pr2392 (optimum = 378032)

Inst	Mean	Min	Median	Midrange	Stdev	Range	IQR	Skew	Dist
pr2392	384764	382226	384820	384840	1028.2	5228.0	1358.5	0.256	0.01796
a10%	385038	383294	385119	385133	1060.0	3679.0	2008.0	-0.146	0.01875
a20%	**384685**	**381697**	384924	384430	1352.4	5466.0	1794.2	-0.406	0.01823
a30%	385381	382452	385416	385315	1266.5	5727.0	1644.2	-0.055	0.01953
a40%	**384398**	382415	383965	385300	1448.2	5770.0	1513.5	1.040	0.01570

Table 4 Statistics for **Time** on 30 executions on pr2392

Inst	Mean	Min	Median	Midrange	Stdev	Range	IQR	Skew
pr2392	59.6647	58.7900	59.7605	59.3965	0.3183	1.2130	0.423	-1.133
a10%	**59.4054**	57.8600	59.5970	58.9190	0.5380	2.1180	0.757	-1.459
a20%	**59.2279**	52.3740	59.4830	56.1875	1.3311	7.6270	0.651	-4.754
a30%	**58.7093**	51.0590	59.4590	55.5215	2.1343	8.9250	0.846	-2.704
a40%	**59.5957**	58.8600	59.6645	59.4300	0.3235	1.1400	0.507	-0.723

The interpretation that we give for the good behavior of the application a20% (which is, in this experiment, better than that of a10%) could stand exactly in a better exploration capacity induced by the moderate "shake" of the edges (through the modified pheromone values). Furthermore, it appears that the applications with higher amplitudes (a30%, a40%) seem to dissipate the exploration efforts. However, the solutions provided by them have quite good characteristics, as the corresponding mean values from Tables 3 and 4 show, when compared to those for the original application (a0%). Once more, the results of the executions on pr2392 prove the reliability of the ant-based application. The last column of Table 3 (Dist), computed as (Mean—optimum)/optimum, is smaller than 2% in all cases.

It needs to be noticed that the analysis presented above is available for the test data gathered during our experiment on pr2392. Tables 3 and 4 might not necessarily display the same pattern (the best behavior for a20%) if other samples are considered for the two random variables (Best and Time).

Concluding, the ACO solving paradigm can be safely applied to complex TSP variants, with dynamicity and uncertainty characteristics embedded at a reasonable level. As our results show that the application can effectively handle uncertain data, it follows that the pheromone matrix can be approximated. This idea can exploit the lack of computing resources in case of expensive optimization. There are intensive efforts for efficient approximation methods (an example for evolutionary methods is [74]). In our case, the total processor load was 50% on average, so the computing system effortlessly solved any instance. The ant-based system was modified according to certain features that the input data were expected to manifest. This practically-driven, hybrid application proved to be resilient and stable, and it delivered very good quality solutions on large instances of the TSP. It is worth mentioning that the application does comply with the recommendations in [75] for the efficiency of the solution space search, namely: useful information is still exchanged during the cooperation phase and diversity is induced exactly by means of the alterations.

6 Conclusions and Future Work

The resilience of a software application package is one important Software Engineering metric, heavily weighting the decisions in military logistics, civil engineering, space industry and defense, broadly speaking when the failure is not an affordable option. The systems that exhibit resilience are more effective than the other ones, as in real life it is hard to predict all the possible situations. This paper investigates the recovery capacity of an application when applied to a rich problem, with multiple dimensions of complexity. Our work shows that the resilience of a biological system (namely, the ant colony) is translated to the artificial system, i.e., the MMAS application is able to overcome the repeated, uncertain data modifications and to produce very good solutions.

This investigation encourages the use of the approach discussed here, as it combines a heuristic with a simple technique that allows the user to deal with a reasonable degree of uncertainty and dynamicity. As a future direction, we plan to search for the thresholds that trigger an unstable behavior, and to study the instability manifested in such situations. Our work can be taken further by considering other complexity dimensions (vertex elimination), other instances from [38] (asymmetric), other TSP problems (Generalized TSP, Vehicle Routing Problem), or other publicly available (under appropriate license) metaheuristic implementations. We believe that the study of the quality of the applications results, when solving uncertain and/or dynamic problems, could benefit from embedding uncertainty and/or dynamicity in the solutions' construction itself. As introduced in [76], this can be done by allowing ACO parameters to vary over time. We intend to explore this direction related to population variance, which was reported to lead to improved robustness.

Acknowledgments G.C.C. and E.N. acknowledge the support of the project "Bacau and Lugano— Teaching Informatics for a Sustainable Society", co-financed by Switzerland through the Swiss-Romanian Cooperation Programme to reduce economic and social disparities within the enlarged European Union.

Annex: Experimental Results

Table 5 presents the detailed results of our experiments.

Table 5 Detailed experimental results for ACOTSP

Inst	Opt	RS	Best	Worst	Av B	Av I	Av T	SD B	SD I
d1291	50801	0.9262	**50801**	50870	50817.6	452.8	30.80	20.59	257.57
a10%		0.9730	**50801**	50877	50822.1	557.4	37.25	20.70	350.76
a20%		0.8533	**50801**	**50840**	**50812.5**	360.7	**24.46**	**13.51**	209.06
a30%		0.9281	**50801**	50877	50831.9	533.9	34.82	28.97	**206.96**
a40%		0.9480	**50801**	50851	50817.8	483.1	24.58	14.34	238.68
d1655	62128	3.8241	62180	62401	62282.5	624.9	56.31	83.07	58.58
a10%		3.7706	62197	62402	62302.5	632.1	55.52	**59.09**	**42.01**
a20%		3.4234	62158	62452	62278.2	**591.3**	**52.37**	96.21	67.89
a30%		3.7091	**62151**	**62376**	62263.5	646.9	56.23	77.48	50.30
a40%		3.8991	62166	62380	**62240.7**	623.0	55.95	67.64	88.85
d2103	80450	1.7333	80512	**80594**	**80560.8**	479.5	**50.86**	**31.29**	60.98
a10%		2.0889	80493	80718	80604.6	**473.8**	50.96	65.45	71.51
a20%		2.1209	80502	80750	80598.0	518.3	54.24	76.69	50.83
a30%		1.7742	**80472**	80691	80562.2	521.4	53.88	76.72	**49.29**
a40%		1.9362	80497	80676	80583.7	503.3	52.61	67.86	58.30
fl1400	20127	0.9213	20203	20366	20255.1	427.5	45.83	54.42	170.44
a10%		1.0549	**20174**	20318	20250.4	502.6	53.61	44.91	**65.51**
a20%		1.0476	20216	20308	20251.3	**414.4**	48.76	**30.42**	114.68
a30%		0.7294	20213	20333	20260.8	368.9	**40.30**	40.70	197.14
a40%		0.7033	20196	**20302**	**20243.3**	440.4	47.54	40.48	94.74
fl1577	22249	0.9380	22352	**22399**	22371.4	710.3	52.93	**15.04**	112.47
a10%		0.9535	**22303**	22406	22360.9	588.5	44.84	32.44	155.33
a20%		1.0565	22309	22431	**22359.3**	648.8	50.88	36.49	**104.49**
a30%		0.9535	22319	22448	22369.8	575.9	**43.38**	41.43	209.52
a40%		0.8957	22330	22408	22369.4	**499.8**	42.65	27.58	152.27
fl3795	28772	0.5000	29581	30097	29861.0	**70.9**	**28.20**	164.70	41.30
a10%		0.6000	29641	**29938**	**29858.8**	83.8	32.95	**94.98**	**34.92**
a20%		0.6500	**29492**	30246	29949.0	110.4	41.80	207.60	42.36
a30%		0.6500	29583	30041	29859.3	94.6	35.56	149.64	51.33
a40%		0.8261	29587	30133	29863.9	125.8	43.54	182.86	78.68
fnl4461*	182566	1.2500	187794	**188162**	188003.6	81.9	86.15	137.16	32.35
a10%		1.6000	**187782**	188381	188096.4	82.3	88.39	200.03	31.53
a20%		1.3684	187802	188322	188065.9	103.7	104.70	168.30	**15.30**
a30%		0.8421	187808	188352	188092.2	76.3	**77.32**	173.18	30.03
a40%		0.8125	187981	188372	188183.2	**75.7**	89.67	**136.82**	20.61
nrw1379	56638	5.3434	56718	56904	**56793.7**	587.8	58.16	58.16	**18.02**
a10%		5.5700	56730	56928	56806.1	584.6	58.00	60.65	37.38
a20%		5.6458	56763	**56846**	56806.0	572.1	**57.55**	**29.66**	33.80
a30%		6.1556	56794	56928	56843.4	**558.2**	58.79	45.75	25.82
a40%		5.7849	56768	56868	56801.9	563.7	58.32	33.69	36.85
pcb1173	56892	2.6111	56906	57115	56959.3	538.6	37.87	63.81	168.12
a10%		2.5809	56893	57161	56990.7	505.5	38.86	76.41	148.91
a20%		2.9722	56905	**57080**	56991.6	592.3	41.78	**57.61**	168.57
a30%		2.2603	56901	57258	57015.9	539.1	**37.33**	131.47	185.44
a40%		2.8195	**56892**	57123	**56957.6**	526.0	40.40	87.07	**107.49**
pcb3038	137694	1.8000	**141608**	**142055**	**141823.2**	136.5	51.35	148.47	36.07
a10%		1.3500	141653	142072	141866.5	136.7	49.20	137.86	30.07
a20%		1.6000	141627	142083	141829.8	144.0	53.64	135.29	18.26
a30%		1.6000	141658	142120	141872.1	146.9	54.57	**126.89**	**11.52**
a40%		1.3889	141632	142308	142015.0	**109.1**	49.49	197.09	29.44
pr1002	259045	2.9873	259081	260492	259703.2	716.2	45.13	440.23	137.39
a10%		2.8861	**259045**	260041	259562.9	821.7	51.28	391.89	128.32
a20%		2.8065	**259045**	**259665**	259393.3	**712.7**	46.16	**223.60**	**124.32**
a30%		2.8012	**259045**	260317	259740.4	716.8	**44.72**	378.56	153.76
a40%		3.0544	**259045**	259977	259557.2	793.7	52.78	336.49	151.99

(continued)

Table 5 (continued)

pr2392	378032	2.7805	**382226**	386514	384986.8	290.7	59.75	1263.39	**6.52**
a10%		2.6341	383331	**386362**	**384626.9**	284.5	59.27	1092.09	9.25
a20%		2.4750	384201	387163	385395.8	268.8	58.79	**977.95**	21.02
a30%		2.3659	383304	386517	385101.0	270.1	57.57	1003.70	34.13
a40%		1.9500	384156	387097	385767.2	**250.4**	**55.71**	1007.76	42.49
rl1304	252948	1.7126	**252948**	253577	253234.3	397.1	45.50	259.15	87.82
a10%		1.7159	**252948**	253675	253295.5	383.8	43.65	263.65	81.93
a20%		1.3523	**252948**	253675	253379.9	406.8	46.48	256.57	**70.41**
a30%		1.3750	**252948**	253849	253479.8	**278.9**	**32.60**	328.51	74.07
a40%		1.5269	**252948**	**253577**	253275.9	397.1	42.36	**248.03**	98.87
rl1323	270199	1.8684	**270199**	**270916**	270718.7	**377.0**	47.44	218.71	56.58
a10%		2.0122	270356	270944	270679.0	408.9	48.23	196.03	63.79
a20%		1.7531	270227	271033	**270677.7**	431.0	50.13	239.50	**35.94**
a30%		2.1875	270466	270932	270704.5	391.8	48.58	**162.74**	51.16
a40%		1.7778	270443	271177	270866.6	392.9	**42.73**	194.77	89.33
rl1889	316536	1.1250	321078	**322656**	321993.5	193.2	51.58	**556.98**	**38.04**
a10%		0.9118	320746	323334	321845.3	**161.8**	**42.01**	709.15	53.12
a20%		0.8235	**318437**	323639	321434.4	162.1	**42.01**	1514.71	54.22
a30%		1.3667	320266	323334	321471.7	179.0	48.01	757.60	46.84
a40%		0.8649	320897	323392	321940.2	192.1	47.89	755.78	45.08
rl5915*	565530	0.6000	583131	585602	584266.2	63.6	79.92	918.85	24.40
a10%		0.6000	583221	586158	584649.0	63.7	85.35	798.79	**16.85**
a20%		0.9000	**580675**	**584967**	**583350.6**	67.1	85.28	1292.42	27.61
a30%		0.7000	583286	585488	584132.4	**61.9**	**76.16**	**706.49**	19.05
a40%		1.3000	582257	585615	584063.2	76.4	89.63	929.14	21.76
rl5934*	556045	0.7000	572479	**576341**	574629.7	60.8	**77.58**	**1248.71**	19.75
a10%		1.0000	572534	576476	574827.9	62.9	84.51	1250.79	18.43
a20%		1.0000	573296	582718	576170.0	**59.2**	85.20	2624.64	23.72
a30%		1.2000	**571738**	576931	574753.8	72.4	90.43	1436.51	**18.34**
a40%		1.2000	572305	577166	574866.2	71.5	84.68	1455.56	21.44
u1060	224094	2.0781	224573	225451	224947.6	732.0	55.41	285.14	**70.29**
a10%		2.1034	224510	225630	224945.2	**617.0**	**51.85**	363.21	73.53
a20%		2.0698	**224254**	225204	**224721.8**	686.7	52.31	293.71	79.51
a30%		2.0887	224403	225476	224887.7	682.5	53.63	315.52	127.68
a40%		2.2248	224468	**225198**	224783.4	705.3	52.63	**249.62**	116.00
u1432	152970	2.2216	153439	153802	153567.2	785.0	42.76	111.50	236.88
a10%		2.5476	153231	153678	153502.2	748.3	46.54	113.02	**154.27**
a20%		2.3207	**153223**	153827	**153484.7**	800.8	43.13	198.78	262.17
a30%		2.5030	153283	153778	153534.0	**652.5**	**39.46**	155.04	176.00
a40%		2.3065	153353	**153654**	153501.4	784.4	41.87	**108.85**	224.49
u1817	57201	2.9709	57294	57619	57402.0	**481.2**	**47.50**	90.65	76.03
a10%		3.3235	**57269**	57512	57395.3	586.1	55.66	65.29	73.82
a20%		3.0463	57272	**57413**	**57349.5**	570.2	53.14	**49.05**	**67.43**
a30%		3.5347	57278	57557	57382.4	532.0	52.85	87.10	91.16
a40%		3.1892	57318	57485	57388.7	582.3	52.41	62.63	101.21
u2152	64253	5.5730	**64324**	64757	64523.5	533.2	58.23	113.18	**26.32**
a10%		5.3974	64393	64714	64549.2	**476.8**	58.02	91.63	44.22
a20%		5.2935	64367	64741	64509.5	520.4	**55.65**	118.61	59.26
a30%		5.6667	64406	64750	64565.4	481.0	57.74	104.30	41.57
a40%		5.1236	64355	**64606**	**64498.1**	532.5	57.71	**73.45**	40.82
u2319	234256	3.6098	235421	235749	235498.3	503.1	58.27	**105.96**	32.78
a10%		3.6753	235421	236611	235682.9	**465.8**	59.00	345.53	37.92
a20%		4.0833	235257	236012	235528.1	503.9	**58.06**	221.13	42.55
a30%		3.8514	235421	236137	235727.4	466.6	58.94	245.56	38.33
a40%		3.8929	**235110**	**235726**	**235454.6**	517.7	58.88	185.60	**31.23**
vm1084	239297	2.4848	239321	239670	239434.8	464.5	46.19	116.91	131.37

(continued)

Table 5 (continued)

a10%		2.3558	**239297**	239572	239422.1	465.8	44.27	95.40	161.96
a20%		2.1391	239321	239670	239489.1	414.0	**36.76**	138.30	106.99
a30%		2.1327	**239297**	239637	239469.0	368.9	33.53	127.12	**74.24**
a40%		2.2843	239303	**239458**	**239400.0**	**409.0**	41.15	**75.00**	141.26
vm1748	336556	1.6000	341159	342667	341906.2	179.5	49.24	556.37	45.38
a10%		1.3871	340352	342492	341684.7	183.7	49.55	575.18	**25.83**
a20%		1.2250	**340346**	342451	**341243.6**	188.0	**45.34**	760.73	54.18
a30%		1.2162	340634	343156	341685.3	**181.2**	46.92	642.61	35.72
a40%		1.3250	340371	**342162**	341486.7	198.9	49.68	**514.81**	26.76
average	191887	2.1733	194389	195282	194855.8	411.9	52.62	294.27	81.06
a10%		2.2122	194388	195350	194859.5	411.5	53.70	297.67	79.63
a20%		2.1678	**194227**	195564	**194836.2**	410.3	53.69	406.73	**77.27**
a30%		2.1913	194353	**195259**	194858.6	**391.5**	**51.39**	305.98	83.31
a40%		2.1681	194382	195321	194871.9	410.6	53.12	**293.87**	85.78

References

1. Dantzig, G.: Linear programming under uncertainty. Manage. Sci. **1**(3–4), 197–206 (1955)
2. Charnes, A., Cooper, W.: Chance-constrained programming. Manage. Sci. **6**(1), 73–79 (1959)
3. Zadeh, L.A.: Fuzzy sets. Inf. Control **8**, 338–353 (1965)
4. Soyster, A.: Convex programming with set-inclusive constraints and applications to inexact linear programming. Oper. Res. **21**, 1154–1157 (1973)
5. Gould, F.J.: Proximate linear programming: an experimental study of a modified simplex algorithm for solving programs with inexact data, University of North Carolina at Chapel Hill, Institute of Statistics Mimeo Series No. 789 (1971)
6. Ben-Tal, A., Nemirovski, A.: Robust solutions of linear programming problems contaminated with uncertain data. Math. Program. **88**, 411–424 (2000)
7. Ben-Tal, A., Nemirovski, A.: Robust convex optimization. Math. Oper. Res. **23**, 769–805 (1998)
8. El-Ghaoui, L., Lebret, H.: Robust solutions to least-square problems to uncertain data matrices. SIAM J. Matrix Anal. Appl. **18**, 1035–1064 (1997)
9. Crişan, G.C., Pintea, C.M., Chira, C.: Risk assessment for incoherent data. Environ. Eng. Manag. J. **11**(12), 2169–2174 (2012)
10. Crainic, T.G., Crişan, G.C., Gendreau, M., Lahrichi, N., Rei, W.: A concurrent evolutionary approach for rich combinatorial optimization. In: Proceedings of the 11th Annual Conference Companion on Genetic and Evolutionary Computation Conference: Late Breaking Papers 2017–2022 (2009)
11. Tlili, T., Krichen, S., Faiz, S.: Simulated annealing-based decision support system for routing problems. In: 2014 IEEE International Conference on Systems, Man and Cybernetics (SMC), pp. 2954–2958 (2014)
12. ADOPT: Advanced Decision-support System for Ship Design, Operation and Training. http://ec.europa.eu/research/transport/projects/items/adopt_en.htm
13. Applegate, D., Cook, W.J., Rohe, S.: Chained Lin-Kernighan for large traveling salesman problems. INFORMS J. Comput. **15**(1), 82–92 (2003)
14. NEOS server, http://www.neos-server.org/neos/
15. Stützle, T.: ACOTSP, http://www.aco-metaheuristic.org/aco-code (2004)
16. Cook, W.J.: In Pursuit of the Traveling Salesman: Mathematics at the Limits of Computation. Princeton University Press, Princeton (2012)
17. Dantzig, G.B., Fulkerson, R., Johnson, S.M.: Solution of a large-scale traveling salesman problem. Oper. Res. **2**, 393–410 (1954)
18. Laporte, G.: A short history of the traveling salesman problem. Canada Research Chair in Distribution Management,Centre for Research on Transportation (CRT) and GERAD HEC Montréal, Canada (2006)

19. Applegate, D.L., Bixby, R.E., Chvátal, V., Cook, W.J.: The Traveling Salesman Problem: A Computational Study. Princeton Series in Applied Mathematics. Princeton University Press, Princeton (2011)
20. Karp, R.M.: Reducibility among combinatorial problems. In: Miller, R.E., Thatcher J.W. (eds.) Complexity of Computer Computations. The IBM Research Symposia, pp. 85–103. Plenum. Press, New York (1972)
21. Applegate, D., Bixby, R., Chvátal, V., Cook, W.: On the solution of traveling salesman problems. Documenta Mathematica, Extra volume ICM 1998(III), 645–656 (1998)
22. Sangalli, A.: Why sales reps pose a hard problem. New Sci. 24–28 (1992)
23. Moscato, P., Buriol, L., Cotta, C.: On the analysis of data derived from mitochondrial DNA distance matrices: Kolmogorov and a traveling salesman give their opinion. In: Pedal, C.D. (ed.) Advances in Nature Inspired Computation: The PPSN VII Workshops, University of Reading, pp. 37–38 (2002)
24. Climer, S., Zhang, W.: Take a walk and cluster genes: a TSP-based approach to optimal rearrangement clustering. In: 21st International Conference on Machine Learning (ICML'04), pp. 169–176, Banff, Canada (2004)
25. Adleman, L.M.: Molecular computation of solutions to combinatorial problems. Science **266**(5187), 1021–1024 (1994)
26. Adleman L.M.: Steering the future of computing. Nature **440**(7083), 383 (2006)
27. Held, M., Karp, R.M.: The traveling-salesman problem and minimum spanning trees. Oper. Res. **18**, 1138–1162 (1970)
28. Held, M., Karp, R.M.: The traveling-salesman problem and minimum spanning trees: part II. Math. Program. **1**, 6–25 (1971)
29. Concorde TSP solver, http://www.math.uwaterloo.ca/tsp/concorde/
30. Google TSP solver, https://code.google.com/p/google-maps-tsp-solver/
31. LKH TSP solver, http://www.akira.ruc.dk/~keld/research/LKH
32. Christofides, N.: Worst-case analysis of a new heuristic for the travelling salesman problem. Technical report 388, Graduate School of Industrial Administration, Carnegie Mellon University (1976)
33. Lin, S., Kernighan, B.W.: An effective heuristic algorithm for the traveling-salesman problem. Oper. Res. **21**(2), 498–516 (1973)
34. Gamboa, D., Rego, C., Glover, F.: Data structures and ejection chains for solving large scale traveling salesman problems. Eur. J. Oper. Res. **160**(1), 154–171 (2005)
35. Braschi, B.: Solving the traveling salesman problem using the simulated annealing on a hypercube. In: Proceedings of the Fourth Conference on Hypercubes, Concurrent Computers and Applications, pp. 765–768 (1989)
36. Dorigo, M., Gambardella, L.M.: Ant colonies for the traveling salesman problem. Technical Report TR/IRIDIA/1996-3 Université Libre de Bruxelles (1996)
37. Benchmark instances for the Travelling Salesman Problem with Time Windows, http://iridia. ulb.ac.be/~manuel/tsptw-instances
38. Library of various sample TSP and TSP-related instances, https://www.iwr.uni-heidelberg.de/ groups/comopt/software/TSPLIB95/
39. GPS-TSP instance, http://cadredidactice.ub.ro/ceraselacrisan/cercetare/ (2015)
40. Jaillet, P., Qi, J., Sim, M.: Routing optimization with deadlines under uncertainty. Working paper, MIT & NUS, http://www.mit.edu/~jaillet/general/publications.html (2014)
41. Jaillet, P.: A priori solution of a traveling salesman problem in which a random set of the customers are visited. Oper. Res. **36**(6), 929–936 (1988)
42. Berman, O., Simchi-Levi, D.: Finding the optimal a priori tour and location of traveling salesman with nonhomogeneous customers. Transp. Sci. **22**(2), 148–154 (1988)
43. Laporte, G., Louveaux, F., Mercure, H.: A priori optimization of the traveling salesman problem. Oper. Res. **42**(3), 543–549 (1994)
44. Rossi, F.A., Gavioli, I.: Aspects of heuristic methods in traveling salesman problem. In: Andreatta, G., Mason, F., Serafini, P. (eds.) Advanced School on Statistics in Combinatorial Optimization, pp. 214–227. World Scientific Publication, Singapore (1987)

45. Bianchi, L., Knowles, J., Bowler, N.: Local search for the traveling salesman problem: correction of the 2-p-opt and 1-shift algorithms. Eur. J. Oper. Res. **162**(1), 206–219 (2005)
46. Bianchi, L., Gambardella, L.M., Dorigo, M.: An ant colony optimization approach to the probabilistic traveling salesman problem. In: 7th International Conference on Parallel Problem Solving from Nature I, vol. 2439, pp. 883–892 (2002)
47. Marinakis, Y., Marinaki, M.: A hybrid honey bees mating optimization algorithm for the probabilistic traveling salesman problem. In: Proceedings of the IEEE Congress on Evolutionary Computation, pp. 1762–1769 (2010)
48. Henchiri, A., Bellalouna, M., Khaznaji, W.: Probabilistic traveling salesman problem: a survey. In: Position papers of the 2014 Federated Conference on Computer Science and Information Systems, Annals of Computer Science and Information Systems, vol. 2, pp. 55–60 (2014)
49. Bianchi, L.: Ant colony optimization and local search for the probabilistic traveling salesman problem: a case study in stochastic combinatorial optimization. Ph.D. Thesis, Université Libre de Bruxelles (2006)
50. Montemanni, R., Barta, J., Gambardella, L.M.: The robust traveling salesman problem with interval data, Technical Report IDSIA 20–05 (2005)
51. Crişan, G.C., Nechita, E.: Solving fuzzy TSP with Ant algorithms. Int. J. Comput., Commun. Control **III**, suppl. issue, 228–231 (2008)
52. Toriello, A., Haskell, W.B., Poremba, M.: A dynamic traveling salesman problem with stochastic arc costs. Oper. Res. **62**, 1107–1125 (2014)
53. Bellman, R.: Dynamic programming treatment of the traveling salesman problem. J. Assoc. Comput. Mach. **9**, 61–63 (1962)
54. de Farias, D.P., van Roy, B.: The linear programming approach to approximate dynamic programming. Oper. Res. **51**, 850–865 (2003)
55. Bertsimas, D.J.: A vehicle routing problem with stochastic demand. Oper. Res. **40**, 574–585 (1992)
56. Secomandi, N.: Analysis of a rollout approach to sequencing problems with stochastic routing applications. J. Heuristics **9**, 321–352 (2003)
57. Cheong, T., White, C.C.: Dynamic traveling salesman problem: value of real-time traffic information. IEEE Trans. Intell. Transp. Syst. **13**, 619–630 (2012)
58. Buhmann, J.M., Mihalák, M., Šrámek, R., Widmayer, P.: Robust optimization in the presence of uncertainty. In: Proceedings of the 4th conference on Innovations in Theoretical Computer Science (ITCS '13), pp. 505–514. ACM, New York, USA (2013)
59. Jaillet, P., Lu, X.: Online traveling salesman problems with service flexibility. Networks **58**, 137–146 (2011)
60. Jaillet, P., Wagner, M.R.: Online vehicle routing problems: a survey. In: Golden, B.L., Raghavan, S., Wasil, E.A. (eds.) The Vehicle Routing Problem: Latest Advances and New Challenges. Springer (2008)
61. MIT Operations Research Center, Laboratory for Information and Decision Systems. https://lids.mit.edu/
62. Schultz, T.R.: In search of Ant ancestors. Proc. Nat. Acad. Sci. **97**(26), 14028–14029 (2000)
63. Wilson, E.O., Hölldobler, B.: Eusociality: origin and consequences. Proc. Nat. Acad. Sci. **102**(38), 13367–13371 (2005)
64. Dorigo, M., Stützle, T.: Ant Colony Optimization. MIT Press, Cambridge (2004)
65. Pintea, C.M.: Advances in Bio-inspired Computing for Combinatorial Optimization Problems. Springer (2014)
66. Crişan, G.C., Pintea, C.M., Pop, P.: On the resilience of an ant-based system in fuzzy environments. An empirical study. In: Proceedings of the 2014 IEEE International Conference on Fuzzy Systems, Beijing, China, pp. 2588–2593 (2014)
67. Nechita, E.: On the Approaches of Classical AI and Embodied AI. Scientific Studies and Research, Series Mathematics and Informatics, vol. 21(1), pp. 175–178 (2011)
68. Nechita, E.: Simulation of Discrete Events Systems (in Romanian). Tehnopress Publishing House, Iaşi (2005)
69. Glover, F., Laguna M.: Tabu Search, Kluwer Academic Publishers (1997)

70. Kirkpatrick, S., Gelatt Jr, C.D., Vecchi, M.P.: Optimization by simulated annealing. Science **220**(4598), 671–680 (1983)
71. Černý, V.: Thermodynamical approach to the traveling salesman problem: an efficient simulation algorithm. J. Optim. Theory Appl. **45**, 41–51 (1985)
72. Birattari, M., Dorigo, M.: How to assess and report the performance of a stochastic algorithm on a benchmark problem: mean or best result on a number of runs? Optim. Lett. **1**(3), 309–311 (2007)
73. Crişan, G.C.: Ant algorithms in artificial intelligence. Ph.D. Thesis, A.I. Cuza University of Iaşi, Romania (2008)
74. Jin, Y.: A comprehensive survey of fitness approximation in evolutionary computation. Soft Comput. **9**(1), 3–12 (2005)
75. Hertz, A., Widmer, M.: Guidelines for the use of meta-heuristics in combinatorial optimization. Eur. J. Oper. Res. **151**, 247–252 (2003)
76. Matthews, D.C., Sutton, A.M., Hains, D., Whitley, L.D.: Improved Robustness through Population Variance in Ant Colony Optimization, Engineering Stochastic Local Search Algorithms. Designing, Implementing and Analyzing Effective Heuristics, Lecture Notes in Computer Science, vol. 5752, pp. 145–149 (2009)

Verifying Compliance for Business Process Logs with a Hybrid Logic Model Checker

Ioan Alfred Letia and Anca Goron

Abstract Given that organizations rely on the support of information systems in automating their business processes, the auditing of these processes is a complex task because it needs to consider both the business process model and the relevant data logs. For the compliance checking of such business situations, we present an extended version of a Hybrid Logics model checking tool, with temporal operators. The support for temporal operators allows for tracing the event logs and the verification of properties, within an abstract representation model corresponding to the intended concern of auditing.

Keywords Compliance checking · Business events logs · Hybrid logics · Model checking · Audit

1 Introduction

Business Process Management (BPM) systems are based on operational processes or sequences of tasks. Considering the flexibility aspect when needing to deal with exceptional situations, the execution of a process must be continuously adapted such that it leads to the desired result. Keeping an event log for the ordering of the process elements in a process instance can provide a certain level of flow control. deLeoni et al. [8] emphasize the importance of verifying event logs traces and their corresponding predefined process model paths in identifying possible deviations and inconsistencies. However, the complexity of today's business processes call for an automated support for audit tasks [14]. Although various (semi-) automated,

I.A. Letia · A. Goron (✉)
Intelligent Systems Group, Department of Computer Science, Technical University of Cluj-Napoca, Baritiu 28, 400391 Cluj-Napoca, Romania
e-mail: anca_d_g@yahoo.com

I.A. Letia
e-mail: Letia@cs.utcluj.ro

© Springer International Publishing Switzerland 2016
I. Hatzilygeroudis et al. (eds.), *Combinations of Intelligent Methods and Applications*, Smart Innovation, Systems and Technologies 46,
DOI 10.1007/978-3-319-26860-6_4

model-based compliance checking approaches have been considered, they mainly focus on the control flow of processes neglecting potentially relevant context based information.

For the compliance checking of business process logs, we propose an extended version of a model checking tool [9], applying the expressivity of Hybrid Logic for verifying properties that are expected to be satisfied within a process execution. Despite the fact that the present literature mentions a plethora of applications using model checking methods [3], the development of viable model checkers for hybrid logic still remains an unexplored field of research [13]. In this paper we introduce the main characteristics of an extended version of the Hybrid Logic Model Checker [9], suggestively named HLMC 2.0, which is able to deal with the expressivity of hybrid multimodal logic along with temporal logic specific operators. By including support for temporal operators, the extended version enables a two-sided perspective in the tracing of event logs by analyzing the changes brought about in time, while depicting relevant data for the captured snapshot of the world. The verification problem is done within a framework based on a Kripke structure that allows for an abstract representation model corresponding to the context of concern for the auditing agent. Hence, in this paper we tackle the problem of Hybrid Logic based model checking [12] and focus on extending the available instrumentation capabilities such that it does not limit the use of all its advantages: from nominals [7], which although behave like propositional variables have one distinctive property and that is that they are true at exactly one state in any model and the interleaving of model logic and temporal logic specific operators.

In the first part of the article we describe shortly main aspects about Dynamic Condition Graphs as a way of visually rendering the ordering of events, Hybrid Logic as specification language for properties to be addressed and the Hybrid Logic Model Checker as a tool support for the task of formal verification. The second part of the article depicts the changes brought to the tool and details the inclusion of temporal operators: Future (F), Past (P), Until (U) and Since (S) for the model checking task. The new implemented version is put at use for the compliance checking task within a suggestive real world scenario. A short presentation of related work and some concluding remarks accompanied by future directions end this paper.

2   Process Mining

Process mining addresses the problem of depicting what is actually happening in a business process [5] by analyzing the structured information encompassed in the so-called event logs [14]. It is a research area combining data mining with process modeling and process analysis in order to improve real processes using techniques such as process discovery, conformance checking, model extension and repair.

Compliance checking refers to verifying given a certain business model, whether the required regulations are met for that model. In a real world, compliance checking has to consider both the predefined behavior for a business model and the actual execution trail for various situations in the running of this business process.

When performing model checking, constraints are represented using formulas usually specified in different types of logic languages such as Linear Temporal Logics (LTL). A search is performed in the model given as a labeled graph to check whether there are states in which the formula holds. Although LTL has been widely used in this direction, the advantages it brings on what it concerns the use of temporal operators is sometimes shadowed by the limitations encountered on what it concerns the knowledge and state based representation of the model. It lacks mechanisms for naming states, for accessing states by names, and for dynamically creating new names for states [9]. Hybrid Logics [12] comes as a solution in this direction as it allows to refer to states in a truly modal framework, mixing elements from first-order logics and modal logics.

3 Instrumentation Support

3.1 Dynamic Condition Response Graphs

A Dynamic Condition Response Graph [10] consists of a set of events, a marking representing the execution state, and four binary relations between the events defining the conditions for the execution of events, the required responses and a novel notion of dynamic inclusion and exclusion of events. Hereto comes a set of actions, a labeling function assigning an action to each event, a set of roles, a set of principals and a relation assigning roles to actions and principals.

Definition 1 [10] A *dynamic condition response graph* is a tuple $G = (\mathsf{E}, \mathsf{M}, \mathsf{Act}, \rightarrow\bullet, \bullet\rightarrow, \pm, l)$ where

1. E is the set of events
2. $\mathsf{M} \in \mathcal{M}(G) = \mathcal{P}(\mathsf{E}) \times \mathcal{P}(\mathsf{E}) \times \mathcal{P}(\mathsf{E})$ is the *marking* and $\mathcal{M}(G)$ is the set of all markings
3. Act is the set of actions
4. $\rightarrow\bullet \subseteq \mathsf{E} \times \mathsf{E}$ is the *condition* relation
5. $\bullet\rightarrow \subseteq \mathsf{E} \times \mathsf{E}$ is the *response* relation
6. $\pm : \mathsf{E} \times \mathsf{E} \rightharpoonup \{+, \%\}$ defines the *dynamic inclusion/exclusion* relations by $e \rightarrow+ e'$ if $\pm(e, e') = +$ and $e \rightarrow\% e'$ if $\pm(e, e') = \%$.
7. $l : \mathsf{E} \rightarrow \mathsf{Act}$ is a labeling function mapping every event to an action.

We let $\mathsf{DCR\ Graphs}$ refer to the model of dynamic condition response graphs.

Fig. 1 Semantics for
temporal logic

$$\mathcal{M}, m \models \top$$
$$\mathcal{M}, m \models p \quad \text{iff} \quad m \in V(p) \;\; \text{for } p \in \text{NOM}$$
$$\mathcal{M}, m \models \neg\varphi \quad \text{iff} \quad \mathcal{M}, m \nvDash \varphi$$
$$\mathcal{M}, m \models \varphi \wedge \psi \quad \text{iff} \quad \mathcal{M}, m \models \varphi \text{ and } \mathcal{M}, m \models \psi$$
$$\mathcal{M}, m \models F\varphi \quad \text{iff} \quad \exists m' \, (Rmm' \wedge \mathcal{M}, m' \models \varphi)$$
$$\mathcal{M}, m \models P\varphi \quad \text{iff} \quad \exists m' \, (Rm'm \wedge \mathcal{M}, m' \models \varphi)$$
$$\mathcal{M}, m \models \psi U\varphi \quad \text{iff} \quad \exists m' \, (Rmm' \wedge \mathcal{M}, m' \models \varphi \wedge$$
$$\forall m'' \, (Rmm'' \wedge Rm''m' \to$$
$$\mathcal{M}, m'' \models \psi))$$
$$\mathcal{M}, m \models \psi S\varphi \quad \text{iff} \quad \exists m' \, (Rm'm \wedge \mathcal{M}, m' \models \varphi \wedge$$
$$\forall m'' \, (Rm'm'' \wedge Rm''m \to$$
$$\mathcal{M}, m'' \models \psi))$$

3.2 Preliminaries of Temporal and Hybrid Logics

Temporal logics (TL) extend propositional logics with temporal operators future
F, past P, until U, since S, so that with the set of propositional symbols PROP $=$
$\{p_1, p_2, \ldots\}$, the syntax of temporal logic is the one below with the semantics in
Fig. 1.

$$\varphi := \top \mid p \mid \neg\varphi \mid \varphi \wedge \varphi \mid F\varphi \mid P\varphi \mid \varphi U\varphi \mid \varphi S\varphi$$

The dual of P is $H\alpha = \neg P\neg\alpha$ and the dual of F is $G\alpha = \neg F\neg\alpha$.

Hybrid logics (HL) extend temporal logics with special symbols that name individual states and access states by name [2]. With nominal symbols NOM $= \{i_1, i_2, \ldots\}$
called *nominals* and SVAR $= \{x_1, x_2, \ldots\}$ called *state variables* the syntax of hybrid
logics is shown below:

$$\varphi := TL \mid i \mid x \mid @_{x_t}\varphi \mid\downarrow x.\varphi \mid \exists x.\varphi$$

With $i \in$ NOM, $x \in$ WVAR, $t \in$ NOM $\cup$ WSYM, the set of *state variables* WSYM $=$
NOM $\cup$ WVAR, the set of *atomic letters* ALET $=$ PROP $\cup$ NOM, and the set of
atoms ATOM $=$ PROP $\cup$ NOM $\cup$ WVAR, the operators @, $\downarrow$, $\exists$ are called *hybrid
operators*.

Figure 2 presents the semantics of Hybrid Logics [9], where $\mathcal{M} = \langle M, R, V \rangle$ is a
Kripke structure, $m \in M$, and g is an assignment.

Fig. 2 Semantics for hybrid
logic

$$\mathcal{M}, g, m \models a \quad \text{iff} \quad m \in [V, g](a), \, a \in \text{ATOM}$$
$$\mathcal{M}, g, m \models @_t\varphi \quad \text{iff} \quad \mathcal{M}, g, m' \models \varphi, \text{ where}$$
$$t \in WSYM$$
$$\mathcal{M}, g, w \models\downarrow x.\varphi \quad \text{iff} \quad \mathcal{M}, g_m^x, w \models \varphi$$
$$\mathcal{M}, g, m \models \exists x.\varphi \quad \text{iff} \quad \text{there is } m' \in \mathcal{M} \text{ such that}$$
$$\mathcal{M}, g_{m'}^x, w \models \varphi$$

4   Illustrative Example

Example 1 We further introduce a running scenario for exemplifying the use of the proposed tool, referring to a claim handling process for a delayed baggage during a flight with multiple connections and operated by different airlines. For space considerations, we consider only a subset of the delayed baggage related regulations specifying that:

- A luggage will be considered delayed if it is delivered to the passenger within more than 6 h from the moment of debarking from the airplane.
- In case that a luggage is reported as delayed, an amount of 5 euros/h will be paid to the complaining passenger up to a maximum of 25 h, which will include also the first 6 h of waiting.

We focus on a passenger named Miguel, traveling from Munich to Bahia Blanca, Argentina, through Sao Paolo and Buenos Aires. When reaching the connecting airport in Buenos Aires, the passenger discovers that his baggage went missing. Hence, a complaint is filled by the passenger requesting the return of his missing luggage and the corresponding compensations. Below we present a short detailing of the passenger itinerary logs from which data was taken and analyzed by the claims responsible personnel:

```
Flight Itinerary
LH504 09Nov Munchen - Sao Paolo:
            20:00 - 05:50 10Nov
 G37682 10Nov Sao Paolo - Buenos Aires:
            13:20 - 15:00 10Nov
 LA4238 10Nov Buenos Aires - Bahia Blanca
            17:45 - 19:05 10Nov
....
Complaint No. G3221
Location: Buenos Aires (Newbery), Argentina
Time and date: 15:55 10Nov
Passenger Name: Miguel Fernandez
```

5   Classical Approach on Hybrid Logic Based Model Checking

Model checking is the problem of determining for a particular model whether a given formula holds in that model [7]. It is used for checking either properties of linear models given as formulas expressed in the language of different types of logics such as Modal or Temporal Logics and the combination of both known as Hybrid Logics [7]. Model checking is based on the use of a Kripke structure for the system modeling.

Fig. 3 Kripke structure
representation of the flight
itinerary

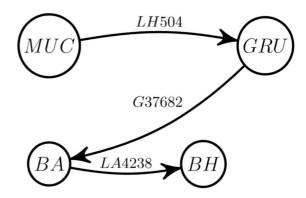

5.1 Model Checking for HL

Definition 2 [7]. A hybrid Kripke structure $\mathcal{M}$ consists of an infinite sequence of states $m1, m2, \ldots$ and a valuation function V that maps ordinary propositions and nominals to the set of states in which they hold, i.e. $\mathcal{M} = \langle \langle m1, m2 \ldots \rangle, V \rangle$

Going back to our example, we further model the travel itinerary as a Kripke structure (Fig. 3). We consider the airports as states in a graph connected by arrows labeled by the performed flights (Fig. 3). Each state is defined by one of the nominals {Munich (MUC), Sao Paolo (GRU), Buenos Aires (BA), Bahia Blanca (BH)} representing the name of the airports. To refer to contextual data highlighting the luggage handling status, we will consider the following two attributes as describing each state:

- l_a (luggage arrival status) which receives the value of *True* if the luggage arrives at the airport and *False* otherwise,
- l_d (luggage departure status), which receives the value of *True* if the luggage is successfully transferred to the next flight and departs from that airport and *False* otherwise.

Table 1 summarizes the luggage handling in each traversed airport. For space considerations, we leave out the exception cases and further assume that, in a flight with multiple connections, once a luggage is confirmed to have had departed from an airport, then it must arrive in the immediate following airport.

Table 1 Luggage handling process from departure airport to final destination

Airport	Luggage departure status	Luggage arrival status
MUC	l_a = False	l_d = True
GRU	l_a = True	l_d = Unknown
BA	l_a = False	l_d = False
BH	l_a = False	l_d = False

Once the business process execution flow is represented as a Kripke structure, the properties to be analyzed are formalized using Hybrid Logic and the formal verification is performed with the support of a model checking tool, which automates the model checking task.

5.2 Tool Support for Compliance Checking

The model checking problem for HL has been widely analyzed in [9]. Explicit model checking algorithms, such as MCLITE and MCFULL, are described for various fragments of HL, representing straight-forward extensions of known algorithms for temporal or propositional dynamic logics and achieving various results between polynomial and exponential time [12].

HLMC[1] is a model checker for hybrid logics. It is a C implementation of algorithms MCLite and MCFull [9]. The tool contains also a parser for the input formula and for the XML based representation of the model. HLMC is a global model checker, reporting to the user all the states of the model given as a Kripke structure in which a certain property is true. HLMC supports as input complex formulas with both hybrid logic operators such as the downarrow binder $\downarrow$, @, the existential operator $\exists$, for all A, but also temporal operators such as future F, past P and since S (for more information on the tool, please refer to [9]). Although the temporal operators are mentioned in [9], the tool does not offer support for their use. For more implementation details on the HLMC tool, we refer the reader to [9]. Considering the limitations of the available tool, we argue the importance of temporal operators for addressing both contextual related information together with the ordering of event occurrences.

6 On Extending HLMC

In [9] a model checker algorithm MCLITE is presented for HL(@, F, P, U, S). We will further consider as supported also the next operator N, which is a special case of the future operator F.

The implementation of MCLITE algorithm was preserved as presented in [9], together with the main notations: the hybrid model $\mathcal{M} = \langle M, R, V \rangle$, the assignment g, and the hybrid formula φ. After termination, every state in the model is labeled with the subformulas of φ that hold at that point.

The algorithm as presented in [9], uses a bottom-up strategy: it examines the subformulas of φ in increasing order with respect to their lengths, from simple formulas to more complicated ones, until φ itself has been checked. The following auxiliary notations from [9] were also used: R to represent an accessibility relation and R^- the inverse of R, while $\mathcal{M}^- = \langle M, R^-, V \rangle$ the inverse of model $\mathcal{M}$. The length of a

---

[1] Available at http://luigidragone.com/software/hybrid-logics-model-checker/.

formula φ, defined by the number of operators (boolean, temporal and hybrid) of φ plus the number of atoms (propositions, nominals and variables) of φ was denoted by $|\varphi|$, while $sub(\varphi)$ was used to denote the set of subformulas of φ exacly like in [9], where $|sub(\varphi)| = |\varphi|$. The model checker MCLITE uses a table L of size $|\varphi| \times |M|$ whose elements are bits.

In the beginning, $L(\alpha, w) = 1 \iff \alpha$ is an atomic letter in $sub(\varphi)$ such that $w \in V(\alpha)$. When MCLITE terminates, $L(\alpha, w) = 1 \iff \mathfrak{M}, g, w \models \alpha$ for every $\alpha \in sub(\varphi)$. Given $\alpha \in sub(\varphi)$ and $w \in M$, we denote by $L(\alpha)$ the set of states $v \in M$ such that $L(\alpha, v) = 1$ and by $L(w)$ the set formulas $\beta \in sub(\varphi)$, such that $L(\beta, w) = 1$ [9].

MCLITE as presented by Franceschet uses three subroutines named MC_F, MC_U and $MC_@$ to check subformulas of the form Fα, αUβ and $@_t\alpha$, past temporal operators being handled by using the inverse model $\mathfrak{M}^-$. To these, we included MC_N to check formulas of the form Nα. Nevertheless, we added two modified versions of the MC_F and MC_U subroutines, implementing the transitive closure operators F+ and U+ and allow the checking of special cases of Kripke structures containing cycles. The corresponding pseudocodes are presented as follows.

6.1 Implementing the Next Operator

The definition of the next operator Nφ states that φ has to hold at the next state [9]. In order to determine N(a), or the states that have as next state the state a, we check the inverse accessibility relation R$^-$ between the node of interest, in this case a, with all the other nodes of the structure (Fig. 4). Those nodes for which the relation holds, are returned as valid states.

Algorithm 1.1 $MC_N(\mathfrak{M}, \alpha)$

```
for  w ∈ L(α)  do
  for  v ∈ R⁻(w)  do
  L(Nα, w) ← 1
  end for
end for
```

Going back to the lost luggage claim scenario, a use of the Next operator would be to determine the last stop before the luggage was discovered as missing. In the old version of HLMC, the following formula is used for the verification process: $< LH > -BA$, which returns airport Sao Paolo as result, while for HLMC 2.0 the same result is returned using the *Next* operator:

$$q1 : Next(BA), \tag{1}$$

which queries for all states that have Buenos Aires airport as immediate successor state. This example is rather simple, however, the utility of the *Next* operator will be better reflected when having to deal with more complex formulas.

Fig. 4 Next operator
explained

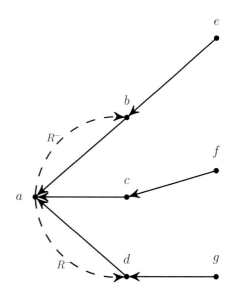

6.2 Implementing the Future Operator

The definition of the Future operator Fφ states that φ has to hold eventually some-where on the subsequent path. Therefore, one must allow an n-depth visiting of predecessor states. The idea was to use an auxiliary list to store all nodes that can reach w obtained by using a backward 1-depth visiting approach.

The correctness of the implementation of the Future operator was tested based on the condition stating that $\exists x.@xF^+x$ holds if and only if the model contains a cycle [9].

The pseudocode is presented below:

Algorithm 1.2 $MC_F(\mathcal{M},\alpha)$

```
for  w ∈ L(α)  do
    auxiliaryList  ←  auxiliaryList + w
end for
while  auxiliaryList
    for  v ∈ R⁻(w)  do
        if  L(Fα, v) = 0
            auxiliaryList  ←  auxiliaryList + v
            L(Fα, v)  ←  1
        end if
    end for
end while
```

When wanting to determine all the states that have as future state the state a as presented in the figure, the MC_N procedure is applied for each node of the model, and when the accessibility relation R$^-$ between the node of interest and a certain other node holds, the MC_N procedure is applied for the other node. This process repeats

Fig. 5 Future operator
explained

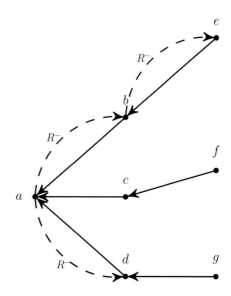

until R^- along the path does not hold for any other nodes (Fig. 5). All the nodes along the path for which R^- holds are returned as valid states. The implementation of the *Past* operator is done in the same way as for the *Future* operator, the only difference being the use of the direct accessibility relation R, instead of the inverse accessibility relation.

To exemplify the use of the *Future* operator, we consider again the lost luggage scenario. If one wants to query the model for the airports from which the luggage successfully departed, one can do that using the *Future* operator:

$$q2 : \exists F(la) \tag{2}$$

Formula $q2$ states that there are states for which the successor states contain airports in which the luggage is reported as arrived. For the presented example, Munich is returned as the only valid result.

6.3 Implementing the Until Operator

$\phi U\varphi$ states that φ has to hold at the current or a future position and ϕ has to hold until that position. At that position ϕ does not have to hold anymore.

To implement the Until operator (U), an auxiliary list was used exactly like in the case of the Future operator to store the visited nodes. However, we observed that performing a one-way visiting of nodes did not suffice as some valid nodes might be left out, while invalid nodes can be wrongly passing as valid. This is why we

Fig. 6 Until operator
explained

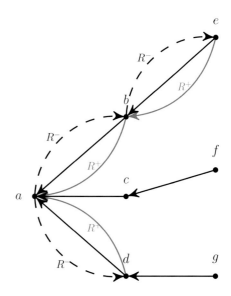

considered using a 2-directions approach. For this, two additional lists were included:
plus[*w*], respectively *minus*[*w*] to store the intermediate results or markings obtained
by considering a forward direction first in the visiting of nodes (R), respectively a
backward direction. The following markings were used: 0:*false*, 1:*true*, 2: or *maybe*.

When determining if a formula of type (*e*)*until*(*a*) holds, one checks the R accessibility relation between the node (*e*) for our case here and all the other nodes
(*a*, *b*, *c*, *d*, *f*, *g*). If the relation holds, the found node is then checked against all
the other nodes. All the valid nodes are stored in a separate list. Then, the same
process is applied, but this time for the inverse accessibility relation R and starting from node (*a*). The final nodes are obtained by performing a *XOR* operation
on the results obtained for each direction (Fig. 6). The following decision logic was
considered:

1. *maybe* + *true* → *true*
2. *maybe* + *false* → *false*
3. *maybe* + *maybe* → *false*
4. *true* + *false* → *false*

The pseudocode is presented as follows:

Algorithm 1.3 $MC_U(\mathcal{M},\alpha,\beta)$

```
for w ∈ M do
    plus[w]  ← 0
    minus[w]  ← 0
end for
```

```
for w ∈ L(α) do
  auxiliaryList ← auxiliaryList + w
end for
for w ∈ L(α) ∪ L(β) do
  for v ∈ L(β) do
    aux1 ← 0
    aux2 ← 0
    for v ∈ R(w) do
      if v ∈ L(α) then
        aux1 ← 1
        if v ∈ L(β) then
          aux2 ← 1
          if !plus[v] then
            auxiliaryList ← auxiliaryList + v
          end if
        end if
      end if
    end for
  end for
  if (aux1 ∧ aux2) then
    plus[w] ← 1
    else if (/(aux1 ∨ aux2)) then
      plus[w] ← 2
      else
      do nothing;
    end if
end for

for w ∈ L(β) do
  auxiliaryList ← auxiliaryList + w
  end for

for w ∈ L(α) do
  aux1 ← 0
  aux2 ← 0
  for v ∈ R⁻¹(w) do
    if v ∈ L(α) then
    aux1 ← 1
    if v ∈ L(β) then
    aux2 ← 1
    if !plus[v] then
    auxiliaryList ← auxiliaryList + v
    end if
    end if
    end if
  end for
  if (aux1 ∧ aux2) then
    minus[w] ← 1
    else if (!(aux1 ∨ aux2)) then
      minus[w] ← 2
      else
      do nothing;
    end if
```

end for

for w ∈ M do
   if $(plus[w] = 1 \land minus[w] = 1)$
   and $(plus[w] \neq 2 \lor minus[w] \neq 2)$ **then**
    L(α U β, w) ← 1
   end if
end for

The *Since* operator is implemented in a similar manner, the only difference being the interchange between the interval limits.

The obtained complexity results for the three procedures are $O = n^2 + m$, for MC_N and MC_F and of $O = n^2\dot{m}$ for MC_U, where $|n|$, respectively $|m|$ represent the number of worlds (M), respectively, of modalities (R) for the Kripke model $\mathcal{M} = \langle M, R, V \rangle$. Comparing our results with the analysis presented by Franceschet et al. [9], optimizations must be performed, but we consider them as acceptable for now for formal verifications done on models with small states sets. However, to be able to apply the model checking task for more complex processes, further improvements are required, which currently represent the subject of our future work.

To exemplify the use of the *Since* operator, we consider a specific case in our scenario when one needs to zoom over a certain event flow, for example, the one which includes the occurrence of the lost luggage. This can be done rather easily in HLMC 2.0 by using the HL formula:

$$q_3 : Bx(@x((ld)Until(!la))), \qquad (3)$$

which must return all states x (labeled using the binder B) in which l_d (luggage departed status) holds and that have as successor state one in which l_a (luggage arrival status) does not hold. The returned states are: Munich and Sao Paolo.

One can observe that by combining hybrid logic operators, such as @ or the down-narrow binder B [15], with temporal operators, we can check ahead the consequences of certain events and look for alternatives from that perspective. Moreover, we can offer a way of verifying sequences of events in an easier manner, by focusing only on certain states of interest.

7 Experiments

Example 2 We extend the initial example with a predefined model for the claim handling process, regarded as a three-stage process:

1. Claim intake: collect all required data, classify the type of claim and routing it to the appropriate claim administrator
2. Claim analysis: an investigation is performed in which all contextual data is analyzed and verified if it complies to the general polices stipulated in the regulations

3. Settlement or rejection of claim: a decision is taken and the client is notified about
   the result.

In order to be able to analyze the execution trail of such a process and compare
it with the predefined model, we argue that a visual support would be necessary to
assist the audit responsible in the task of depicting possible deviations from the initial
flow. While, Kripke structures are already used for capturing the behavior of business
processes within the formal verification task, they do not offer enough information
for the audit agent about events ordering constraints. Considering this aspect, we
complement the classical modeling approach with the representation of constraints
using DCR Graphs (Fig. 7).

Figure 7 highlights that checking-in a luggage is a primary condition ($\rightarrow \bullet$) for
the transfer to occur. As a response ($\bullet \rightarrow$), a check is performed to verify whether

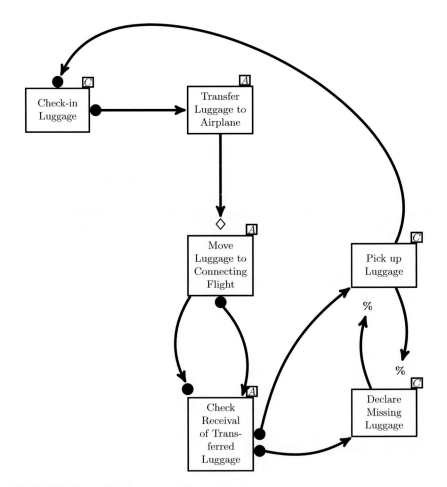

Fig. 7 DCR Graph of the luggage handling process

the luggage was sent to the following flight. The acknowledgment of receiving the luggage represents a condition for the passenger to be able to pick up his luggage at the destination, while a negative response indicates that the luggage might be missing. The exclude relation ($\rightarrow\%$) between the pick-up and luggage missing events shows that they cannot occur at the same time. The notation A is used for referring to airport personnel in charge of a certain task, while P for referring to the Passenger. Considering the ordering of events (Fig. 7) and the available logs (Fig. 3), we further verify with the support of HLMC 2.0 at whether the condition referring to the case that once the luggage was given to the airport personnel it must be transferred to the corresponding airplane unless it is the destination airport, holds for our model:

$$q_4 : \exists Bx(@x(la)Until@x(la\&!ld)) \qquad (4)$$

Formula q_3 checks whether there exists a state x (specified using the binder B) in which the luggage arrival status attribute holds up to the point where the status changes. Although q_3 should return both MUC and GRU for the normal execution flow, the output contains only airport MUC, signaling a transfer issue in the Sao Paolo airport. In such a case, further investigations are required for verifying the baggage handling procedure performed at airport GRU. We check further the condition that once a luggage successfully departs from an airport, it must arrive in the following airport:

$$q_5 : Bx(By(@x((ld)\&(Next(x)->y))\&@y(la))) \qquad (5)$$

Formula q_4 returns all states x in which the luggage departure status holds and that have as immediate successor a state in which the luggage arrival status is true. Airport Munich is returned as the only valid result, proving again the erroneous luggage handling process within airport GRU. Hence, by analyzing the event logs captured within the Kripke structure and considering the prespecified ordering of events depicted in the DCR Graph based representation, a delimitation of the interval which might contain issues in the execution process can be easily obtained using the model checker, information which further enables, for our case here, an approximation of the delay period to be performed and used in the claim handling process.

Although the above examples are rather simple, they are used as an indicator of the practical advantages offered by the proposed audit approach. Furthermore, we argue that by combining Temporal and Hybrid Logics for the specification and verification of specific event logs or aspects of interest, we obtain a boost of expressivity when it comes to specifying contextual data to be analyzed and allows the verification of complex properties whether referring to functional or control based aspects in a flow of events, or whenever it is needed to take into consideration organizational related and specific pieces of data of interest from the captured logs.

8 Related Work

Reasoning formalisms for model checking. Linear Temporal Logics (LTL) [6] and Computational Tree Logic (CTL) [16] have been widely used for the model checking task [3]. However, the advantages they bring on what it concerns the use of temporal operators are sometimes shadowed by the limitations encountered on what it concerns the knowledge and state based representation of the model. They lack mechanisms for naming states, for accessing states by names, and for dynamically creating new names for states [9]. Hybrid Logics [12] comes as a solution in this direction as it allows to refer to states in a truly modal framework, mixing features from first-order logics and modal logics. Checking business processes for compliance to rules using CTL and LTL [5] is focused mainly on covering the functional and control-flow perspective. Our approach distinguishes itself by the ability to consider also the context related data, based on the expressive power of HL($@, \downarrow, \exists, A, F, Next, P, U, S$). But hybridization is not simply about quantifying over states, but about handling different type of information in a uniform way [4]. This is useful for the audit of business process logs as we deal with different types of information that needs to be verified against a model.

Model visualization methods. Model checking offers the advantage of allowing for violations to be identified and resolved prior to execution. Although, at a first glance, using model checking might be too complex for business experts, it might bring great value in the end. That is why, through the proposed solution, we tried complementing the representation model for the business process logs with a visual representation of the main constraints and properties to be considered as DCR Graphs [10]. Another visualization solution is presented in [11], and it is based on the use of Guard-Stage-Milestone (GSM) meta-model for representing business process execution flows and help non-expert users gain insights into the operational mode of a business process. The temporal perspective is captured through timestamps and snapshots of different parts of the model. However, the complexity of the approach, although it offers a detailed representation of the main aspects of the execution flow, makes it difficult to be applied and understood by business stakeholders. Hence, the chosen visualization approach based on DCR Graphs, although it offers a less refined representation, it allows to capture the main constraints of the operational model, depicting easily the main dependencies between the flow of events.

Comparison with other tools. There are various mentionings of the HLMC based verification in the present literature, as for example the model checking of strategic equilibria in the analysis of game-like systems [15]. Moreover, other automated tools that exploit the xepressivity of hybrid logic are proposed, such as HyLMoC [13]. Similar to HLMC, HyLMoC is a model checker explicitly oriented towards hybrid logic formulas and Kripke models. HyLMoC checks whether a formula is true (globally true or locally true) in a given model, and the model has to be taken as a finite structure. As opossed to HLMC, HyLMoC uses a top-down approach in evaluating formulas, instead of the usual bottom-up, which checks first the leaves of the parse tree and identifies the worlds at which the atomic subformulas are true, then reconstructs the entire modal formula by navigating the inverse of the respective accessibility relations [13]. However, this approach becomes unfeasible when binders are included. HyLMoC implements some parts of the HLMC tool, as for example MCLITE and a version of MCFULL algorithm. Based on the results presented in [13] the bottom-up approach of MCLITE works better than the top-down approach used in HyLMoC, especially when large models are considered. By comparing the two tools, one can notice that, as opposed to HLMC, the current implementation of HyLMoC encounters difficulties mainly regarding the complexity of the formulas analyzed than the increase of formula dimension.

9 Discussion and Perspectives

To the best of our knowledge, the extended version of HLMC presented in this paper is the only tool currently available which is able to perform model checking on hybrid logic formulas combined with temporal operators. The new version was created with the intention to be used as an instrumentation support for a more complex task of conformance checking of dynamic systems governed by rules that change in time [1] and we argue that it can complement existing sets of tools and techniques for risk evaluation, response and monitoring.

This tool is meant to improve the understanding of what happens at the trace level according to the properties that the human actor knows to be significant from the point of view of the runs of the declarative process. The ability of the Hybrid Logic to better express the abstract representations of very complex models in real world scenarios is a plus that can improve the understanding of how processes interact in complex situations.

Improvements can be found in the current implementation method and efforts are oriented towards testing and comparing the performances of the tool on scenarios of different complexity levels. Regarding future directions, we aim to add the ability to take additional data into account. By incorporating knowledge, for example, adding epistemic relations over states to the models and the underlying knowledge operators to the language, we would obtain an expressive logic capable of strategic reasoning, allowing an easier identification of discrepancies between the declarative model and the log. Additionally, selected parts of the event logs could be separated and further used in diagnosing identified deviations from the initial model and determining their severity. Another direction, would be to integrate the new version of the model checker in a framework for self-adaption and repair of Hybrid Logic Models.

The preliminary theoretical results presented in this paper open the way towards a new approach on model checking dynamic business processes, bringing additional value by the interleaving of modeling facilities with formal verification methods, which enabl a sense of control over the operational mode of a business process within the real world.

Acknowledgments The authors gratefully acknowledge the support by CNCSIS-UEFICSU, National Research Council of the Romanian Ministry for Education and Research, project ID_170/ 2009.

References

1. Aldewereld, H., Alvarez-Napagao, S., Dignum, F., Vazquez-Salceda, J.: Making norms concrete. In: 8th International Conference on Autonomous Agents and Multiagent Systems, pp. 807–814 (2009)
2. Areces, C., Ten Cate, B.: Hybrid logics. In: Blackburn, P., Van Benthem, J., Wolter, F. (eds.) Handbook of Modal Logic, pp. 821–868. Elsevier, Amsterdam (2007)
3. Arellano, G., Argil, J., Azpeitia, E., Benítez, M., Carrillo, M., Góngora, P., Rosenblueth, D.A., Alvarez-Buylla, E.R.: "antelope": a hybrid-logic model checker for branching-time boolean grn analysis. BMC Bioinform. **12**(1) (2011)
4. Blackburn, P., Tzakova, M.: Hybrid languages and temporal logic. Logic J. IGPL **7**(1), 27–54 (1999)
5. Caron, F., Vanthienen, J., Baesens, B.: Comprehensive rule-based compliance checking and risk management with process mining. Decis. Support Syst. **54**(3), 1357–1369 (2013)
6. Clarke, E.M., Grumberg, O., Hamaguchi, K.: Another look at ltl model checking. Formal Methods Syst. Des. **10**(1), 47–71 (1997)
7. Cranefield, S., Winikoff, M.: Verifying social expectations by model checking truncated paths. J. Logic Comput. **21**(6), 1217–1256 (2011)
8. de Leoni, M., Maggi, F.M., van der Aalst, W.M.: An alignment-based framework to check the conformance of declarative process models and to preprocess event-log data. Information Systems (2014)
9. Franceschet, M., de Rijke, M.: Model checking hybrid logics (with an application to semistructured data). J. Appl. Logic **4**, 279–304 (2006)
10. Hildebrandt, T.T., Mukkamala, R.R.: Declarative event-based workflow as distributed dynamic condition response graphs, Springer. In: Programming Language Approaches to Concurrency and Communication-cEntric Software (2010)
11. Hull, R., Damaggio, E., De Masellis, R., Fournier, F., Gupta, M., Heath, F.T., III, Hobson, S., Linehan, M., Maradugu, S., Nigam, A., Sukaviriya, P.N., Vaculin, R.: Business artifacts with guard-stage-milestone lifecycles: managing artifact interactions with conditions and events. In: Proceedings of the 5th ACM International Conference on Distributed Event-Based System, DEBS'11, pp. 51–62, ACM, New York, NY, USA (2011)
12. Lange, M.: Model checking for hybrid logic. J. Logic, Lang. Inform. **18**(4), 465–491 (2009)
13. Mosca, A., Manzoni, L., Codecasa, D.: HyLMoC: a model checker for hybrid logic. In: Gavanelli, M., Riguzzi, F. (eds.) CILC09: 24-esimo Convegno Italiano di Logica Computazionale. Ferrara, Italy (2009)
14. Schultz, M.: Enriching process models for business process compliance checking in erp environments. In: vom Brocke, J., Hekkala, R., Ram, S., Rossi, M. (eds.) Design Science at the Intersection of Physical and Virtual Design. Lecture Notes in Computer Science, vol. 7939, pp. 120–135. Springer, Berlin (2013)
15. Troquard, N., Hoek, W., Wooldridge, M.: Model checking strategic equilibria. Chapter model checking and artificial intelligence, pp. 166–188. Springer, Berlin (2009)
16. Zhang, Y., Ding, Y.: Ctl model update for system modifications. J. Artif. Int. Res. **31**(1), 113–155 (2008)

Smarter Electricity and Argumentation Theory

Menelaos Makriyiannis, Tudor Lung, Robert Craven, Francesca Toni
and Jack Kelly

Abstract The current, widespread introduction of smart electricity meters is resulting in large datasets' becoming available, but there is as yet little in the way of
advanced data analytics and visualization tools, or recommendation software for
changes in contracts or user behaviour, which use this data. In this paper we present
an integrated tool which combines the use of abstract argumentation theory with
linear optimization algorithms, to achieve some of these ends.

1 Introduction

Recent research [12] has shown that UK energy prices are rising at eight times the
rate of average earnings, and the industry's trade body, Energy UK, recently warned
that "household bills could rise by another 50 % over the next 6 years" [11]. Partly
as a response to this trend, and partly also in an attempt to improve the sustainability
of current forms of energy usage in at least the medium term, the UK Department of
Energy and Climate Change has required a gradual introduction of 'smart meters', to
be rolled out to all UK homes by 2020 [10]. EU legislation from 2012 states that "80 %
of all electricity meters in the EU have to be replaced by smart meters by 2020" [9].

Smart meters record the utility (e.g., electricity, gas, water) consumption within a
household—our focus here is on electricity. After a regular, and often short interval,
the energy consumed is recorded, and can be stored or sent to a local database for
future analysis. This can result in a substantial amount of data's being available for
analysis. However, there is currently no complete software package available which

M. Makriyiannis · T. Lung · R. Craven (✉) · F. Toni · J. Kelly
Department of Computing, Imperial College London, London SW7 2AZ, UK
e-mail: robert.craven@gmail.com

© Springer International Publishing Switzerland 2016
I. Hatzilygeroudis et al. (eds.), *Combinations of Intelligent Methods
and Applications*, Smart Innovation, Systems and Technologies 46,
DOI 10.1007/978-3-319-26860-6_5

makes an innovative use of the data, to properly empower individual users to control their electricity bills by providing detailed analysis of existing usage, and intelligent recommendations on how to adjust usages or contract providers, in order best to meet the users' demands. Recommendation websites (e.g., www.uswitch.com) do exist, but it is hard for users to be confident that their recommendations are fair, since the detailed reasons for the recommendations made are often not available to users [2]. Such websites also do link to the actual readings from users' smart meters, and no use is made of disaggregated readings of appliances.

In this paper we present preliminary work aimed at providing detailed support, by way of visualization and recommendation, to electricity users. Specifically, we make use of abstract argumentation (AA) theory [4], a branch of logical AI, to construct theoretically-underpinned arguments for how users might change electricity contracts and behaviour in order to minimize their electricity bill. The use of AA is combined with linear optimization algorithms, and both are applied to data output from smart meters currently available (these included Current Transformers[1] and EDF Transmitter Plugs[2] We use two large, real-world data sets for testing and evaluation. The resulting web-based implementation is available online, as a prototype, at http://smartelectricity.io.[3]

The paper is organized as follows. In Sect. 2 we present background on abstract argumentation, and describe the data sets used. In Sect. 3 we describe central parts of the functionality of our tool. In Sect. 4 we present results from the detailed evaluation, and comparison, made. We conclude in Sect. 5.

2 Background

Abstract argumentation [4] is used to represent and resolve the relations of conflict between opposing arguments, without delving into the internal logical structure of those arguments.

Definition 1 An *abstract argumentation framework* is a tuple $(Args, \rightsquigarrow)$, where

- $Args$ is a set of *arguments*;
- $\rightsquigarrow \subseteq Args \times Args$ is the *attack relation*.   ⌟

For example, the set *Args* might represent arguments for changing electricity contracts to a different provider based on current or projected electricity usage patterns, with the attack relation $\rightsquigarrow$ being between arguments for different contracts.

---

[1] See http://www.elkor.net/pdfs/AN0305-Current_Transformers.pdf.

[2] See https://shop.edfenergy.com/Item.aspx?id=540&CategoryID=1.

[3] Log in using username *houseX*, password *houseX*, for $X \in \{1, 2, 3, 4\}$.

A total set of 'winning' arguments can be determined in several alternative ways, including the following.

Definition 2 Where $(Args, \rightsquigarrow)$ is an AA-framework:

- $A \subseteq Args$ is *conflict-free* if there are no $a, b \in A$ such that $a \rightsquigarrow b$;
- $A \subseteq Args$ is *admissible* if (i) it is conflict-free, and (ii) for any $a \in A$ such that $b \rightsquigarrow a$, there is $a' \in A$ such that $a' \rightsquigarrow b$;
- $A \subseteq Args$ is *preferred* if it is maximally (w.r.t. $\subseteq$) admissible.   ⌙

Admissibility of a set of arguments thus ensures for them a form of internal consistency, and means they collectively defend themselves from external attack; being preferred requires this in a $\subseteq$-maximal way. (Other semantics have been widely studied, but play no role in this paper. See [1] for details.)

We were able to make use of two large electricity smart-meter datasets in testing and evaluating our implementation.

First, the *UK Power Data* (UKPD) dataset [6][4] contains detailed power consumption data from four London houses recorded over several months. The data were recorded using two types of sensor, Current Transformers[5] and EDF Transmitter Plugs,[6] for individual appliance measurements. The data contains entries for individual appliances, as well as for the entire house consumption.

Secondly, the *Household Electricity Survey* (HES) dataset [3] includes data from 250 owner-occupied households from the UK, from 2010 to 2011. 26 households were monitored for a full year, and the remaining 224 for 1 month, on a rolling basis throughout the trial. This data is disaggregated—that is, appliance-specific—with no aggregates provided per house.

In addition to the two datasets, we collected and represented electricity contract data. Widely-used contracts from two of the main UK electricity suppliers (British Gas[7] and EDF Energy[8]) were used.

3 Functionality

3.1 Data Representations

Contract data was standardized to be represented in a UML format of the form

---

[4]http://www.doc.ic.ac.uk/~dk3810/data/.

[5]http://www.elkor.net/pdfs/AN0305-Current_Transformers.pdf.

[6]https://shop.edfenergy.com/Item.aspx?id=540&CategoryID=1.

[7]http://www.britishgas.co.uk/.

[8]http://www.edfenergy.com/.

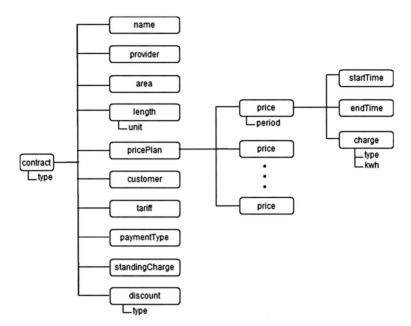

A configurable, stand-alone application was written to perform data standardization for the smart electricity readings from the two datasets described in Sect. 2. This standardization had several aspects. The data in each of the two sets represents the electricity usage over a given interval of n seconds; but the value of n varies, both between datasets and also within specifically the HES dataset—as determined by the particular smart meters in use. The UKPD data uses an interval of 6 seconds, and the HES data has an interval of 2 min for the majority of houses, and 10 min for the others. In the case of the UKPD set, this would mean 14,400 data points per house per day—a substantial storage cost. As our implementation is intended to be the prototype for a web-based application, data calls with such amounts of data would hinder the user experience and also make computation much more expensive, both on the server-side and on the client-side. Accordingly, we selected a time of 1 hour to be the standard length of interval, and aggregated both datasets appropriately.—Other aspects of the standardization were less significant. We chose the DateTime format[9] for the time-stamp, and since the HES dataset only exists in a disaggregated form (values per appliance), we also summed the data to produce total series of values per household—values which are used in some of the functions of our implementation.

3.2 Data Visualization

For the data visualization components of the implementation, we used the HighchartsJS[10] framework, which allows for a wide variety of visualization types, and is

---

[9]http://msdn.microsoft.com/en-gb/library/system.datetime.aspx.

[10]http://www.highcharts.com/.

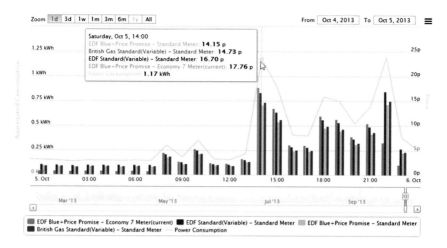

Fig. 1 Aggregated consumption and cost chart

supplied with extensive documentation. In the present subsection we describe four of the data visualizations we developed.

First, we discuss the **aggregated consumption and cost chart**. (For a screenshot, see Fig. 1.). This chart represents two types of data: a line series depicts the consumption of electricity over a given period, with points at each hourly interval as determined by the data standardization discussed in Sect. 3.1. Secondly, at each point of the aggregated consumption data line, the respective cost of that consumption sample under one or more contracts is visualised in the form of column bars. This may be the cost under the contract the user is currently under or any alternative contracts, or a combination thereof. The user may choose the total period for which the data is displayed, either using a number of hard-coded choices, or by sliding a bar to fix the period precisely; and the contracts whose pricing is displayed can be toggled on or off in the legend. Finally, as the user hovers over points on the graph, the precise values of the nearest data point are displayed. This form of visualization enables the user to see, of their aggregated electricity consumption, how a selection of contracts compare in a fine-grained way.

The aggregated consumption and cost chart may prove useful to a user if the user wants to examine their usage as a whole, with a view to changing contract. However, the user may wish, instead, to alter their behaviour whilst remaining with their current provider and on their current contract. For this, visualizations using the disaggregated datasets are more appropriate. Thus, secondly, we implemented a **disaggregated consumption chart** (see Fig. 2). As with the aggregated consumption and cost chart, the user may select a period for which the data is visualized. A number of appliances are selected, and the visualization then presents the changing total consumption, per hourly interval, over the period selected, as well as the electricity consumption per individual appliance. This enables the user to see frequent spikes

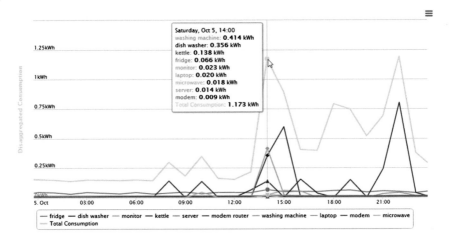

Fig. 2 Disaggregated consumption chart

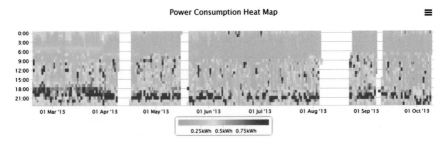

Fig. 3 Power consumption heat map

in the power consumed by particular appliances—thus allowing them to consider altering their behaviour with respect to those devices which use the most power.

The **heat map consumption chart**, shown in Fig. 3, gives the user a snapshot of their entire consumption history in a single glance, irrespective of prices for any contract. On the vertical axis are times of the day, and on the horizontal axis points during the period for which the consumption is shown. Colours are then used to depict varying levels of power usage; higher values are shown in red, and lower in green, with yellow intermediate. The heat map can be useful to a consumer because it allows her or him to immediately identify patterns in consumption which might need to be addressed, i.e., periods when consistently high levels of power are used.

Finally among the visualizations we select to show here, our **average consumption appliance chart** (shown in Fig. 4) depicts the average electricity consumption over the entirety of a selected data interval, at a single glance. The consumption of the ten appliances which use the most electricity in the given period is shown, together with a section ('other appliances') for those appliances whose electricity consumption is known but are not shown in the ten. Since the datasets we use sometimes show

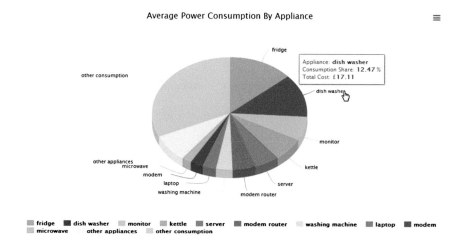

Fig. 4 Average consumption appliance chart

a discrepancy between the sum of the values for the disaggregated appliances, and the aggregated consumption value (since not all appliances are included in the disaggregation), there is also a sector in the pie chart for 'other consumption', representing the other appliance usage not included in the disaggregated data.

3.3 Contract Comparison

In order to give a summary of comparisons between different contracts of electricity providers, we also allow a detailed tabular comparison between the prices a user would have been charged, on a set of different contracts, for selectable, different periods of their historical power usage (see Fig. 5). This **Contract Summary Table** provides a breakdown of how a user's electricity would have been charged according to different contracts, over a number of different periods; the entries are ordered according to the saving, or loss, to be accrued for that period, with the row for the current contract highlighted. For example, for the user whose electricity is being analyzed in Fig. 5, a switch to EDF's 'Blue + Price Promise' Economy 7 contract is indicated: this would involve a saving of £37.88 for the selected period. Switching contracts to either EDF's 'Blue + Price Freeeze' Economy 7 contract, or British Gas's 'Fix and Control' Economy 7, would have resulted in the user being £12.82 or £19.47 out of pocket, respectively.

In addition, a **contract price comparison** visualization, as shown in Fig. 6, displays part of the information shown in the Contract Summary Table (Fig. 5), but with the possibility of splitting the prices for different contracts according to the various periods during the day for which there are different rates. For example, in Fig. 6 two

Contract Name	Selected Interval (£)	Standing Charges (£)	Entire Consumption (£)	Past Year (£)	Past 6 Months (£)	Past 3 Months (£)	Past Month (£)	Switching Savings (£)
EDF Blue+Price Promise - Economy 7 Meter	260.14	45.56	410.16	399.88	204.67	111.14	38.61	37.88
EDF Standard(Variable) - Economy 7 Meter	290.70	45.56	458.52	447.01	228.86	124.49	43.28	7.32
British Gas Standard(Variable) - Economy 7 Meter *(current)*	298.02	62.68	469.57	457.86	234.32	126.67	43.91	0.00
EDF Blue+Price Freeeeze - Economy 7 Meter	310.83	45.56	490.40	478.06	244.81	133.29	46.35	-12.82
British Gas Fix and Control - Economy 7 Meter	317.48	77.99	499.87	487.46	249.32	134.32	46.50	-19.47

Fig. 5 Contract summary table

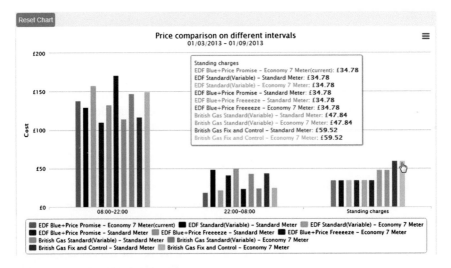

Fig. 6 Contract Price Comparison Chart

different periods are shown (08:00–22:00 and 22:00–08:00), as well as the cost of standing charges for the various contracts shown. Which contracts are shown can be varied, and the periods depicted can be merged to give aggregated costs.

Although the Contract Summary Table and Contract Price Comparison Chart are useful for visualising contracts and giving aggregate cost figures calculated on the basis of the user's consumption, they do not explicitly state comparison conclusions and justification. Further, they are passive features and their purpose is to complement comparisons rather than being the source of them. Thus, we also supplement the previously mentioned comparisons with **Contract comparison recommendations**, which give detailed, argumentation-theoretic justifications for the results summarized in the table, by constructing an AA framework and determining its preferred sets of arguments. In the remainder of this section we describe this construction and the resulting recommendations.

Let X and Y be two contracts to be compared. Each contract splits the day into a number of intervals $X_1, \ldots, X_m$ and $Y_1, \ldots, Y_n$. These may not overlap, so that there is a set $\mathsf{T}_{X,Y}$ of $k \leqslant m \times n$ *tariff-stretches*:

$$\mathsf{T}_{X,Y} = \{X_i \cap Y_j \mid X_i \in \{X_1, \ldots, X_m\},$$
$$Y_j \in \{Y_1, \ldots, Y_n\}, X_i \cap Y_j \neq \emptyset\}$$

Each tariff-stretch represents a period of time over which each contract's tariff is constant, and for each tariff-stretch $T \in \mathsf{T}_{X,Y}$ either the total charge for contract X, or Y, is greater, or they are equal. Let $X(T)$ be the amount charged by contract X for tariff-stretch $T \in \mathsf{T}_{X,Y}$, and $Y(T)$ the amount charged by contract Y for T; since contracts also have standing charges, let $X(S)$ be the standing charge for X, and $Y(S)$ that for Y. Further, let $X(\{T_1, \ldots, T_n\})$ be $X(T_1) + \cdots + X(T_n)$, etc. Define $w : \mathsf{T}_{X,Y} \cup \{S\} \to \{X, Y, \{X, Y\}\}$ so that:

$$w(U) = \begin{cases} X & \text{if } X(U) < Y(U), \\ Y & \text{if } X(U) > Y(U), \\ \{X, Y\} & \text{if } X(U) = Y(U). \end{cases}$$

so that $w(U)$ gives the 'winner' between contracts X and Y, in the sense of the contract which charges least for that stretch or for the standing charge.

So, given two contracts X and Y, the resulting abstract argumentation framework $(Args_{X,Y}, \leadsto_{X,Y})$ has:

$$Args_{X,Y} = \{(\{U_1, \ldots, U_k\}, C) \mid U_1, \ldots, U_k \in \mathsf{T}_{X,Y} \cup \{S\},$$
$$w(U_1) = \cdots = w(U_k) = C, (C = X \vee C = Y)\}$$
$$\leadsto_{X,Y} = \{((T_1, C_1), (T_2, C_2)) \mid C_1 \neq C_2,$$
$$C_1(T_1) \leqslant C_2(T_2)\}$$

We then find sets of arguments of the resulting argumentation framework, as illustrated next.

Example 1 Consider two contracts, X and Y:

- X charges electricity at a rate of 15 p/kWh between 8 a.m. and 9 p.m., and 13.54 p/kWh between 9 p.m. and 8 a.m.; and there is a standing charge of £40;
- Y charges electricity at a rate of 16.54 p/kWh between 9 a.m. and 10 p.m., and 13.54 p/kWH between 10 p.m. and 9 a.m., with a standing charge of £35.

This gives $\mathsf{T}_{X,Y}$ and standing charges (S) as follows (shown with the kWh used for sample periods, and the resultant prices for X and Y):

	p/kWh (X/Y)	kWh	Price (X/Y), £	w
T_1, 08–09	15/13.54	166	24.9/22.48	Y
T_2, 09–21	15/16.54	260	39/43	X
T_3, 21–22	14.54/16.54	300	43.62/49.62	X
T_4, 22–08	14.54/13.54	400	58.16/54.16	Y
S	n/a	n/a	40/35	Y

This gives the AA-framework shown below. (The arguments have been annotated with the £ value of the saving made on the winning contract; so, for example, $(\{T_1\}, Y) : 2.4236$ represents that over tariff-stretch T_1, contract Y is the cheaper contract by £2.4236.)

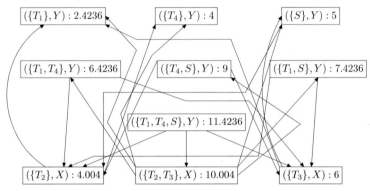

The preferred set of arguments in this example has all and only arguments for Y. ⌟

The argumentation framework exhibits a number of features. If there is a 'winning' contract, then there will be at least one argument in $Args_{X,Y}$ supporting that contract which is unattacked by any other argument; in fact, any unattacked argument in the framework represents a decisive argument, based on cost, for the contract it supports. Further, there may be two preferred sets of arguments; this represents the case where, although for particular tariff-stretches one contract may be better than another, overall, the contracts would require equal payments for the period and usage under consideration.

The output from the argumentation component is processed in order to give a prose output as a recommendation to the user. For example, w.r.t. the example above:

> Even though in the stretch 09:00–22:00 contract X is less costly by £10.00, contract Y in the interval 22:00–09:00 is less costly by £6.42 and has £5 less standing charges, which is enough to make it less costly overall by £1.42, over the selected date interval.

Contract X wins by £10.00 over some tariff-stretches, but over the remaining tariff-stretches and the standing charge, Y wins by 11.42. Thus Y wins overall. Whilst the sum-total of the difference between the charges for the two contracts might easily be computed without any argumentation theory, the use of an argumentation-theoretic underpinning brings the logical relations between the structure of charges to the surface, and provides a stepping-stone which lets the natural-language rendition be easily computed.

3.4   Recommendation Generation

Recommendation generation is a central feature of our approach. The recommendations we implemented are divided into two broad categories, of **General** and
Appliance recommendations. In the present subsection we select some of the many
recommendation capacities that our suite of tools provides.

   General recommendations concern a user's aggregated consumption (without
paying attention to any appliance-specific, disaggregated data). We have already
mentioned, in Sect. 3.3, contract comparisons: if all contracts available are compared, these can be used to provide a **Contract switch recommendation** based in a
straightforward fashion on the user's history of electricity usage. If the user wishes
to stay with her or his current electricity provider and contract, then a **Consumption
behaviour recommender** can provide advice on how the cost of the user's bill might
have been lowered by shifting portions of the aggregate power consumption between
what, in Sect. 3.3, were called 'tariff-stretches'. The Consumption behaviour recommender finds portions of the power usage which occur near the boundaries between
tariff-stretches, and optimizes the price of the bill by allocating this power to a different stretch. The tool does this by looking at the total usage for a given period (again,
selectable by the user).

   A **Reduced usage estimator** allows the user to see what percentage of the various
tariff-stretches' usage needs to be cut if a desired reduction in the price of a bill is to
be achieved. One sliding bar represents the cost of electricity, and more sliding bars
(one each per tariff-stretch) show the percentage of usage allocated to that stretch.
When the user manipulates the slider on any one bar, the others compensate in real
time. If the total bill cost is reduced, aggregate electricity usage is shifted between
tariff-stretches to indicate how the user must change behaviour; and the process
also works in reverse, as the user alters the amount of power usage allocated to any
particular stretch.

   Shifted usage recommendations, perhaps the most useful of the general recommendations provided, combine the Contract switch recommendations with Consumption behaviour recommendations. The recommendations examine minimal
changes to the user's aggregate electricity usage patterns (up to a threshold which
is easily configurable) to give overall advice on the best bill price possible. This is
done with respect to all possible contracts with all possible electricity providers. The
presumption here is that a user may be able to shit some of their usage of electricity
either side of a boundary in tariff-stretches. A sample such recommendation is shown
in Fig. 7.

   Appliance recommendations use disaggregated data to achieve more fine-
grained recommendations. A basic **Appliance consumption table** shows the average
hourly consumption per applicance, together with the total price per appliance over
a user-selectable interval. An **Appliance recommendation table** performs a similar function to the Consumption behaviour recommender described above, but for
individual pieces of equipment. This tool calculates the n appliances, a portion of
whose usage might mostly profitably be switched from one tariff-stretch to another,

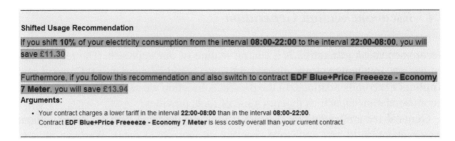

Fig. 7 Shifted usage recommendation

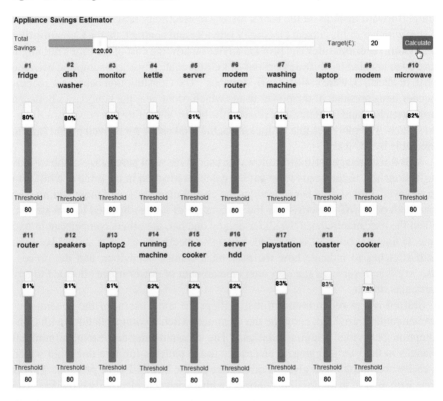

Fig. 8 Appliance savings estimator

displays this information to the user, and finds the resultant monetary benefit. An **Appliance savings estimator** (Fig. 8) allows the user to see the effect of different reductions in the usage of individual appliances on total bill cost whilst remaining with the current provider and contract, *or* to set a desired bill reduction and see how this can best be balanced by reducing appliances' use. Thresholds can be set—a useful constraint when a subset of appliances may only be reduced to a certain degree. Recommendations are displayed as percentage values of current usage. **Additional**

To achieve this, you may consider

- using your **monitor** only **2** times a day, instead of your usual **3** times a day.
- using your **kettle** only **2** times a day, instead of your usual **3** times a day.
- using your **laptop** only **2** times a day, instead of your usual **3** times a day.
- using your **microwave** only **2** times a day, instead of your usual **3** times a day.
- using your **speakers** only **2** times a day, instead of your usual **3** times a day.
- using your **laptop2** only **1** time a day, instead of your usual **2** times a day.
- using your **running machine** only **1** time a day, instead of your usual **2** times a day.
- using your **rice cooker** only **1** time a day, instead of your usual **2** times a day.
- using your **playstation** only **1** time a day, instead of your usual **2** times a day.
- using your **toaster** only **1** time a day, instead of your usual **2** times a day.
- using your **cooker** only **1** time a day, instead of your usual **2** times a day.

Fig. 9 Additional appliance recommendations

recommendations for appliances (Fig. 9) converts these percentages into cardinal values. This works by estimating, on by analysis of appliance power usage values, the number of times a given appliance is used per day on average, and calculating how many times fewer the appliances should be used to give the desired cost reduction.

Most of the recommendation functionality is implemented as the solution of linear optimization problems solved in real time; we omit the details, which are straightforward. They can be found in [7, 8].

4 Evaluation

We wanted to provide some basic testing of the various forms of recommendation our tool offers. The ideal way to do this would have been to roll out the tool as a beta to users, who could have collected disaggregated data for a year, then followed different recommendations the tool gave. A form of validation could then be found if the users' bills, over the following year, were low relative to all possible combinations of provider and contract.

Unfortunately, this sort of testing was markedly infeasible: we did not have the time to conduct a study of such length, nor the number of users, and did not want to become immersed in technicalities of data collection when our primary intention was with generating recommendations, and the *use* of data. Accordingly, we made use of the existing datasets. Though the HES dataset included 250 houses, only 26 of these were monitored for a year, with the rest of the houses monitored for 1 month only. In order to have a reasonably large enough monitoring period for recommendation evaluation, we therefore confined ourselves to the 26, with the addition of the 4 houses from the UKPD, each of which has data from a period of several months.

Each house was assigned a contract randomly, designated as its current contract. The houses' data was then split into a set from which recommendations were made, and a set against which, once the recommendations had been followed, the effect was measured (call the former *training*, the latter *testing* data). For example, for **Contract switch recommendations**, we split the data in two ways: first, into an 80 %/20 % division of training/testing; and secondly, into a 50 %/50 % division (different splits were used to ensure that any results found were not unduly sensitive to any arbitrariness in the split). Our tool was used to find a recommendation for the best contract to switch to, given the user's power usage for the training set. We then tested whether, if that recommendation were followed and the user switched contracts, the bill would be the lowest possible. Of the 30 datasets we used, our recommendation tool gave the optimal contract switch in 87 % of the cases for the 80/20 split, and 90 % of the cases in the 50/50 split. (The difference indicates a shift in usage pattern in accuracy indicates cases where the usage pattern of the user shifts from the training to the testing data.) The results for the the 80/20 split are shown in Table 1.

Table 1 Results for Contract Switch recommendations (80–20 % split)

#	Training data recommendation (80%)	Testing data recommendation (20%)
1	Remain with the current contract	Remain with the current contract
2	EDF Blue+Price Promise(Standard), £5.74	Remain with the current contract
3	EDF Blue+Price Promise(Economy 7), £12.44	EDF Blue+Price Promise(Standard), £0.13
4	EDF Blue+Price Promise(Standard), £22.16	Remain with the current contract
5	EDF Blue+Price Promise(Standard), £35.31	Remain with the current contract
6	EDF Blue+Price Promise(Standard), £20.84	Remain with the current contract
7	EDF Blue+Price Promise(Standard), £53.64	Remain with the current contract
8	EDF Blue+Price Promise(Economy 7), £66.30	Remain with the current contract
9	EDF Blue+Price Promise(Economy 7), £26.76	EDF Blue+Price Promise(Standard), £2.41
10	EDF Blue+Price Promise(Economy 7), £7.07	Remain with the current contract
11	EDF Blue+Price Promise(Standard), £27.69	Remain with the current contract
12	EDF Blue+Price Promise(Economy 7), £43.69	Remain with the current contract
13	EDF Blue+Price Promise(Economy 7), £59.45	EDF Blue+Price Promise(Standard), £3.84
14	EDF Blue+Price Promise(Economy 7), £48.74	Remain with the current contract
15	EDF Blue+Price Promise(Standard), £22.72	Remain with the current contract
16	EDF Blue+Price Promise(Economy 7), £53.37	Remain with the current contract
17	EDF Blue+Price Promise(Economy 7), £1.44	Remain with the current contract
18	EDF Blue+Price Promise(Standard), £4.46	Remain with the current contract
19	EDF Blue+Price Promise(Standard), £18.60	Remain with the current contract
20	EDF Blue+Price Promise(Economy 7), £33.67	Remain with the current contract
21	EDF Blue+Price Promise(Standard), £30.91	Remain with the current contract
22	EDF Blue+Price Promise(Standard), £35.96	Remain with the current contract
23	EDF Blue+Price Promise(Standard), £28.20	Remain with the current contract
24	EDF Blue+Price Promise(Standard), £71.48	Remain with the current contract
25	EDF Blue+Price Promise(Standard), £8.83	Remain with the current contract
26	EDF Blue+Price Promise(Economy 7), £13.24	EDF Blue+Price Promise(Standard), £3.67
27	Remain with the current contract	Remain with the current contract
28	EDF Blue+Price Promise(Standard), £8.97	Remain with the current contract
29	EDF Blue+Price Promise(Standard), £7.56	Remain with the current contract
30	EDF Blue+Price Promise(Economy 7), £12.55	Remain with the current contract

In the table, houses are rows. The left main column represents the recommendation made for which contract to switch to, together with the saving made if that contract was chosen. The right column indicates whether the recommended contract is the cheapest one possible for the testing data—a green cell indicates it is, a red cell that it is not, with the discrepancy in the cost shown.

Now consider the appliance recommendations. These consist of advice to shift the pattern of use for a single appliance to a different time of day, so as to take account of different rates for alternative tariff-stretches in contracts. Only that amount of power consumption which is within an hour of a change in tariff price is considered—thus, if a given contract changes its tariff at 10pm, then with respect to this boundary, only that amount of the usage from either 9pm–10pm (to be shifted an hour later), or 10pm–11pm (to be shifted an hour earlier) is considered. The intention here is to minimize the amount of disruption to the user's pattern of activities, and therefore to make the recommendation more likely to be accepted.

Some sample results for 15 houses are shown in Table 2. As before, we divided the data 50/50 into training and test data. The 'Appliance' column indicates which appliance was selected for its usage to be shifted—we choose the appliance whose usage would, when shifted, give the highest saving with respect to the training data (the saving is shown in the second column, which therefore can be interpreted as a prediction of the saving to be made if the applicance is shifted). The third column indicates the actual saving made when the recommendation is applied to the test data:

Table 2 Results for appliance recommendations

#	Appliance	Training data result (50 %) (£)	Testing data result (50 %) (£)	Difference
1	Kettle	0.73	0.73	0.00
2	Gas, boiler	1.17	0.39	0.78
3	Fridge freezer	0.48	0.46	0.02
4	Freezer upright	0.82	0.71	0.11
5	TVLCD	0.00	0.54	−0.54
6	Tumble dryer	0.63	1.29	−0.66
7	Kettle	0.26	0.14	0.12
8	Hi_Fi	1.12	1.13	−0.01
9	TVLCD	5.48	5.18	0.30
10	Tumble dryer	2.80	4.67	−1.87
11	TVLCD	0.59	0.55	0.04
12	TVLCD	2.30	2.42	−0.12
13	TVLCD	1.10	1.11	−0.01
14	TVLCD	1.37	1.73	−0.36
15	Heater electric portable	0.00	1.48	−1.48
Total	–	18.85	22.53	−3.68

we shifted the quantity of power output in the way the appliance recommendation tool suggested, and noted the saving made. The fourth column shows the difference between the second and third columns, and this is therefore a measure of how well the recommendation tool has performed. The smaller the value here, the more accurate the projected saving has been. (Note that, overall, more savings were made than the training data suggested.)

Other tests indicate a comparable rate of success for the various recommendation components. Detailed results are omitted—see [8] for details.

5 Conclusion

With the advent of ubiquitous electricity smart-meter presence in households, there is a need for advanced software tools which can enable the end-user to understand his or her electricity usage. The availability of large amounts of data should enable users to make informed decisions about how best to choose providers and contracts, and how to alter their own electricity usage to lower their bills, thus saving both money, and also, potentially, energy (for higher costs typically correspond to more energy used).

In the current paper we presented a prototype, web-based implementation underpinned by the use of argumentation theory and linear optimization, as a step towards this end. The tool combines a large number of visualization and recommendation components, and tests indicate it may be highly successful in reducing power bills. Abstract argumentation was used to bring rational structures underpinning recommendations to the user.

Future work will, first, study automatic methods for the disaggregation of appliances from aggregated data. The UKPD dataset we used, which provided both aggregate and disaggregated power usage measurements, was collected by attaching individual smart meters to appliances in a household. While this method is likely to give very accurate readings, it will not be the form of smart metering which is typically installed in UK households. Thus, if those facets of our recommendation tool which depend on data for individual appliances are to be supported, we must use a method for the automatic disaggegation of aggregated readings (see [5] for one approach to this).

Another direction is that of machine learning and user profiling. The large amounts of data obtained from smart meters could be used to devise mathematical models for the consumption patterns of users, leading to more accurate recommendations. This can be associated with a notion of user profiling, which would involve formulating a profile for each particular household that characterises its consumption habits precisely. For example, there might be various types of profile such as 'student', 'family of four', 'young couple' or even completely user-specific profiles. The derived profile can then be used by the system to offer only the most fitting recommendations for that household.

Finally, we are interested in implementing the recommendation and visualization tool as a mobile application. As the application developed is a web-based, it already provides the infrastructure on which a mobile application can be developed to complement it. This would give users more control over their data as well as pave the way for new functionalities. For instance, the mobile device can be used to issue 'live' recommendations. This could indicate to the user in real time that an appliance has passed a consumption threshold, or that the period of the day in which the contract tariff is highest is about to start. Thus, recommendations can be issued in the form of notifications on a mobile device, if and when needed. This could eventually eliminate the time lag between the recommendation generation and its actual implementation, particularly in the case of live data collection.

Acknowledgments Work was funded by EPSRC grant EP/J020915/1 (TRaDAr: Transparent Rational Decisions by Argumentation).

References

1. Baroni, P., Giacomin, M.: Semantics of Abstract Argumentation Systems. In: Rahwan, I., Simari, G.R. (eds.) Argumentation in Artificial Intelligence, pp. 25–44. Springer, Berlin (2009)
2. Canocchi, C.: Price comparison websites 'hide cheapest energy deals' (2014). http://www.thisismoney.co.uk/money/bills/article-2799949/price-comparison-websites-hide-cheapest-energy-deals-claims-rival.html
3. Department for Environment, Food and Rural Affairs: Household electricity study, ev0702 (2009–2012). http://randd.defra.gov.uk/Default.aspx?Module=More&ProjectID=17359
4. Dung, P.M.: On the acceptability of arguments and its fundamental role in nonmonotonic reasoning, logic programming and n-person games. Artif. Intell. **77**(2), 321–358 (1995)
5. Kelly, J.: Disaggregating smart meter readings using device signatures. M.Sc. thesis, Department of Computing, Imperial College London (2011)
6. Kelly, J., Knottenbelt, W.: UK-DALE: a dataset recording UK domestic appliance-level electricity demand and whole-house demand. ArXiv e-prints (2014)
7. Lung, T.: Smart electricity recommendations. M.Sc. thesis, Department of Computing, Imperial College London (2013)
8. Makriyiannis, M.: Smart electricity by argumentation. M.Eng. thesis, Department of Computing, Imperial College London (2014)
9. The European Commission: Energy: Commission paves the way for massive roll-out of smart metering systems (2012). http://ec.europa.eu/energy/gas_electricity/smartgrids/doc/20120309_smart_grids_press_release.pdf
10. The Guardian: Smart energy meters in every uk home by 2020 (2009). http://www.theguardian.com/environment/2009/may/11/smart-meters-energy-efficiency
11. The Guardian: Energy uk warns that household bills could rise by 50 % over six years (2013). http://www.theguardian.com/money/2013/nov/12/energy-uk-warns-50-percent-rise-household-bills
12. The Observer: Energy bills rise by 37 % in three years (2013). http://www.theguardian.com/money/2013/nov/16/energy-prices-rise

Online Argumentation-Based Platform for Recommending Medical Literature

Andrei Mocanu, Xiuyi Fan, Francesca Toni, Matthew Williams
and Jiarong Chen

Abstract In medical practice, choosing the correct treatment is a key problem [1]. In this work, we present an online medical recommendation system, RecoMedic, that selects most relevant medical literature for patients. RecoMedic maintains a medical literature repository in which users can add new articles, query existing articles, compare articles and search articles guided by patient information. RecoMedic uses argumentation to accomplish the article selection. Thus, upon identifying relevant articles, RecoMedic also explains its selection. RecoMedic can be deployed using single-agent as well as multi-agent implementations. The developed system has been experimented with by senior medical Ph.D students from Southern Medical University in China.

1 Introduction

In medical practice, choosing the correct treatment is a key problem. Modern practice has emphasized the role of using explicit evidence to make this decision, and a cornerstone of this evidence is generated from randomized controlled trials (RCT).

A. Mocanu (✉)
University of Craiova, Craiova, Romania
e-mail: mocanu.andrei@ucv.ro

X. Fan · F. Toni · M. Williams
Imperial College London, London, UK
e-mail: xf309@imperial.ac.uk

F. Toni
e-mail: ft@imperial.ac.uk

M. Williams
e-mail: matthew.williams@imperial.ac.uk

J. Chen
Southern Medical University, Guangzhou, China
e-mail: garwingchan@163.com

© Springer International Publishing Switzerland 2016
I. Hatzilygeroudis et al. (eds.), *Combinations of Intelligent Methods and Applications*, Smart Innovation, Systems and Technologies 46,
DOI 10.1007/978-3-319-26860-6_6

These compare two (or more) treatments in a cohort of defined patients who are randomly allocated to each treatment arm, thus minimizing bias. However, some areas of medicine may generate many trials (e.g. there are over 500 new RCTs/year on breast cancer alone), which makes it difficult to identify the optimal treatment for each patient. In such cases, we would like a tool to help us identify the ideal study, which would match each patient characteristic most closely.

Formal argumentation, as a powerful reasoning methodology, has been used extensively in AI in the last two decades (e.g. see [2–4] for an overview). One unique feature of argumentation is that while performing computation as a form of reasoning, argumentation also gives an explanation to the computation. Thus, argumentation can serve as a versatile methodology for applications that need both correct computation as well as transparent explanation.

RecoMedic is a trial recommendation system which uses argumentation to match individual patients to published clinical trials focusing on *brain metastases*. Figure 1 describes RecoMedic's main use case scenario: a patient visits a doctor for medical advice; after some examination, a set of patient characteristics are collected; to determine the most suitable treatment for the patient, the doctor consults RecoMedic with this patient's characteristics specifically. RecoMedic queries its internal medical literature repository to identify the most relevant trial for this particular patient. Since this query is executed using argumentation (Assumption-based Argumentation (ABA) [5] in particular), RecoMedic not only returns the most relevant literature but also an explanation for this recommendation.

Decision making is a process of selecting good *decisions* amongst several alternatives based on the *goals* met by decisions. RecoMedic views clinical trial recommendation as a decision making problem where trials are alternative decisions and patient characteristics are goals. Thus, RecoMedic uses techniques developed in argumentation-based decision making. Decision making with ABA has already

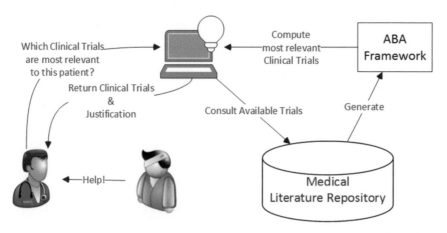

Fig. 1 Recommendations and justifications with argumentation

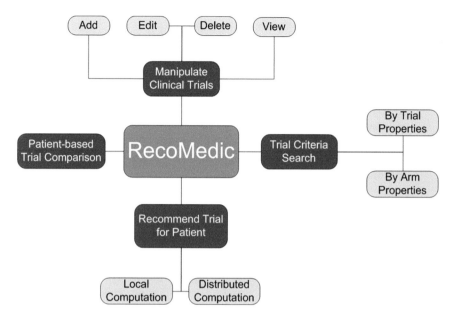

Fig. 2 RecoMedic features

been studied in [6–9]. Our work incorporated it into a platform directly accessible to end-users.

From a functionality point of view, as illustrated in Fig. 2, RecoMedic has four main features: (i) manipulating clinical trials (medical literature), (ii) searching literature in the repository based on certain criteria, (iii) recommending the most relevant literature for given patients, and (iv) comparing medical literature relevance for patients. We describe each of these features in later sections.

The rest of this paper is organized as follows. Section 2 reviews the argumentation-based decision making theory we used throughout this work. Sections 3–6 present the four main features of our system, outlined above. Section 7 describes the system design and our implementation. Section 8 presents the system evaluation conducted with medical literature on brain metastases. Section 9 discusses related works. Section 10 concludes.

2 Background

This work relies upon Decision Frameworks [7] and Assumption-based Argumentation (ABA) [5].

Decision frameworks [7] are tuples $\langle \mathrm{D}, \mathrm{A}, \mathrm{G}, \mathrm{T_{DA}}, \mathrm{T_{GA}}, \mathrm{P} \rangle$ with:

- a (finite) set of decisions $\mathrm{D} = \{d_1, \ldots, d_n\}, n > 0$;

- a (finite) set of attributes $A = \{a_1, \ldots, a_m\}, m > 0$;
- a (finite) set of goals $G = \{g_1, \ldots, g_l\}, l > 0$;
- a partial order over goals, P, representing the preference ranking of goals;
- two tables: T_{DA}, (size $n \times m$), and T_{GA}, (size $l \times m$)[1]:

  - for all $d_i \in D, a_j \in A, T_{DA}[d_i, a_j]$ is either:
    1, representing that d_i *has* a_j, or
    0, representing that d_i does not have a_j, or
    u, representing that it is unknown whether d_i has or does not have a_j;
  - for all $g_k \in G, a_j \in A, T_{GA}[g_k, a_j]$ is either
    1, representing that g_k is *satisfied* by a_j, or
    0, representing that g_k is not satisfied by a_j, or
    u, representing that it is unknown whether g_k is satisfied by a_j or not.

Given a decision framework $DF = \langle D, A, G, T_{DA}, T_{GA} \rangle$, a decision $d_i \in D$ *meets* a goal $g_j \in G$, wrt DF, iff there exists an attribute $a_k \in A$, such that $T_{DA}[d_i, a_k] = 1$ and $T_{GA}[g_j, a_k] = 1$. $\gamma(d)$, where $d \in D$, denotes the *set of goals met by d*.

Given a decision framework $\langle D, A, G, T_{DA}, T_{GA}, P \rangle$ the *most preferred decisions* are the decisions meeting the more preferred goals that no other decisions meet, formally defined as follows. For every $d \in D$, d is *most preferred* iff the following holds for all $d' \in D \setminus \{d\}$:

- for all $g \in G$, if $g \notin \gamma(d)$ and $g \in \gamma(d')$, then there exists $g' \in G$, such that:

  - $g' > g$ in P
  - $g' \in \gamma(d)$, and
  - $g' \notin \gamma(d')$.

Assumption-based Argumentation (ABA) *frameworks* [5] are tuples $\langle \mathcal{L}, \mathcal{R}, \mathcal{A}, \mathcal{C} \rangle$ where

- $\langle \mathcal{L}, \mathcal{R} \rangle$ is a deductive system, with $\mathcal{L}$ the *language* and $\mathcal{R}$ a set of *rules* of the form $\beta_0 \leftarrow \beta_1, \ldots, \beta_m (m \geq 0, \beta_i \in \mathcal{L})$;
- $\mathcal{A} \subseteq \mathcal{L}$ is a (non-empty) set, referred to as *assumptions*;
- $\mathcal{C}$ is a total mapping from $\mathcal{A}$ into $2^{\mathcal{L}} - \{\{\}\}$, where each $\beta \in \mathcal{C}(\alpha)$ is a *contrary* of α, for $\alpha \in \mathcal{A}$.

In an ABA framework, given a rule ρ of the form $\beta_0 \leftarrow \beta_1, \ldots, \beta_m$, β_0 is referred to as the *head* and $\beta_1, \ldots, \beta_m$ as the *body*. We focus on *flat* ABA frameworks, where no assumption is the head of a rule.

In ABA, *arguments* are deductions of claims using rules and supported by sets of assumptions, and *attacks* are directed at the assumptions in the support of arguments. Informally, following [5]:

- *an argument for (the claim)* $\beta \in \mathcal{L}$ supported by $A \subseteq \mathcal{A}$ (denoted $A \vdash \beta$ in short) is a (finite) tree with nodes labeled by sentences in $\mathcal{L}$ or by τ,[2] the root labeled

---

[1] $T[x, y]$ denotes the cell in row labelled x and column labelled y in T.

[2] $\tau \notin \mathcal{L}$ represents "true" and stands for the empty body of rules.

by β, leaves either τ or assumptions in A, and non-leaves β' with, as children, the elements of the body of some rule with head β', or τ if this body is empty;

- *an argument* $A_1 \vdash \beta_1$ attacks an argument $A_2 \vdash \beta_2$ iff β_1 is a contrary of one of the assumptions in A_2.

Attacks between (sets of) arguments in ABA correspond to attacks between sets of assumptions, where *a set of assumptions A attacks a set of assumptions A'* iff an argument supported by a subset of A attacks an argument supported by a subset of A'.

With argument and attack defined for a given $\mathcal{F} = \langle \mathcal{L}, \mathcal{R}, \mathcal{A}, \mathcal{C} \rangle$, standard argumentation semantics can be applied in ABA [5], e.g. *a set of assumptions is admissible* (in $\mathcal{F}$) iff it does not attack itself and it attacks all $A \subseteq \mathcal{A}$ that attack it; *an argument* $A \vdash \beta$ *is admissible (in $\mathcal{F}$) supported by $A' \subseteq \mathcal{A}$* iff $A \subseteq A'$ and A' is admissible (in $\mathcal{F}$); *a sentence is admissible* (in $\mathcal{F}$) iff it is the claim of an argument that is admissible supported (in $\mathcal{F}$) by some $A \subseteq \mathcal{A}$.

As shown in [6, 7, 9], ABA can be used to model decision making problems and compute "good" decisions. The ABA framework for computing the *most preferred decisions* in a decision framework $\langle D, A, G, T_{DA}, T_{GA}, P \rangle$ with $D = \{d_1, \ldots, d_n\}, n > 0, A = \{a_1, \ldots, a_m\}, m > 0, G = \{g_1, \ldots, g_l\}, l > 0$ is defined as $AF = \langle \mathcal{L}, \mathcal{R}, \mathcal{A}, \mathcal{C} \rangle$ for which:

- $\mathcal{R}$ consists of all the following rules[3]:

for all $g_t, g_r \in G$, if $g_t > g_r \in P$ then $prefer(g_t, g_r) \leftarrow$;

for $k = 1 \ldots n$; and $j = 1 \ldots m$, if $T_{DA}[k, i] = 1$ then $hasAttr(d_k, a_i) \leftarrow$;

for $j = 1 \ldots m$; and $i = 1 \ldots l$, if $T_{GA}[j, i] = 1$ then $satBy(g_j, a_i) \leftarrow$;

$met(D, G) \leftarrow hasAttr(D, A), satBy(G, A)$;

$notSel(D) \leftarrow met(D', G), notMet(D, G), notMetBetter(D, D', G)$;

$metBetter(D, D', G) \leftarrow met(D, G'), notMet(D', G'), prefer(G', G)$;

- $\mathcal{A}$ consists of all the following sentences:

for all $d_k \in D, sel(d_k)$;

for all $d_k \in D$ and $g_j \in G, notMet(d_k, g_j)$;

for all $d_k, d_r \in D, d_k \neq d_r$ and $g_j \in G, notMetBetter(d_k, d_r, g_j)$;

- $\mathcal{C}$ is such that:

for all $d_k \in D, \mathcal{C}(sel(d_k)) = \{notSel(d_k)\}$;

for all $d_k \in D$ and $g_j \in G, \mathcal{C}(notMet(d_k, g_j)) = \{met(d_k, g_j)\}$;

for all $d_k, d_r \in D, d_k \neq d_r$ and $g_j \in G$,

$\mathcal{C}(notMetBetter(d_k, d_r, g_j)) = \{metBetter(d_k, d_r, g_j)\}$.

Theorem 1 in [7] sanctions that the aforementioned ABA framework is a sound and complete argumentative computational counterpart for decision making in a way that a decision d is most preferred iff the argument $\{sel(d)\} \vdash sel(d)$ is admissible.

---

[3]We use *schemata* with variables (D, D', A, G, G') to represent compactly all rules that can be obtained by instantiating the variables over the appropriate domains as follows: D, D' are instantiated to decisions, A to attributes, G, G' to goals.

3 Manipulating Clinical Trials

An essential feature of our system is allowing the users to add new clinical trials and thus improve the selection range for each patient-based medical literature recommendation. The input fields are a collection of representative data and metadata for clinical trials on the treatment of brain metastases and are classified in two categories: *Clinical Trial Design* and *Patient Characteristics*. The *Clinical Trial Design* category contains administrative fields such as the `Trial ID`,[4] `PMID`,[5] number of arms, clinical trial phase, recruitment area, and trial start/end year, but it also states medical information such as the eligible age/number of metastases/performance status for patients and any excluded histology or extra-cranial disease they may have. The *Patient Characteristics* category includes information for one or more clinical trial arms such as the `Arm ID`,[6] number of patients enrolled in the clinical trial, performance status range of the patients enrolled in the trial and specific factors such as the percentage of patients with non-small cell lung cancer or the percentage of patients with stable extra-cranial disease.

A major concern while designing the UI, which is presented in Fig. 3, was to minimize the amount of errors made by the users when filling in the web form. Thus, we decided to use interactive elements such as range sliders, dropdown lists, checkboxes, and multiple selection boxes.

In order to also accommodate faster (keyboard-only) input we give users the option to type in associated text fields, but the values entered are carefully limited and validated prior to submission to the server (e.g. the possible values for the eligible number of metastases are integers from 0 to 4 and the plus sign, standing for *more than 4 metastases*). Another way by which we improve the input speed is by allowing the users to modify several arm values at once, since many remain unchanged or there are only small differences from one arm to the next. Moreover, in medical literature we may find cases where one or several of the form fields are missing or not applicable, so the fields can also be disabled from the web form. Since it is possible for the `TrialID` and the `PMID` to be missing as well, we added identifiers which are automatically generated and are guaranteed to be unique and not null.

Our users also have the option to edit or delete the clinical trial information they have previously entered into the system. The clinical trials in the system are visible by all users but are only editable by the person who submitted them. Furthermore, the system provides a way for its users to quickly browse through literature by supplying the abstract and a direct link to the paper, but it also features an embedded PDF viewer for reading the full text version without leaving the platform.

---

[4]Unique identifier for clinical trials.

[5]Unique number assigned to each PubMed record.

[6]Unique identifier for the arm of a clinical trial.

Add Clinical Trial   My Clinical Trials   Criteria Search   Recommend Trial

Clinical Trial Design

Patient Characteristics

MODIFY ALL ? ☑

TrialID ? ☑   N/A

PMIDNumber ? ☑   N/A

Arm 1

NoArms ? ☑   1

ArmID ? ☑   N/A

Phase ? ☑   Study Design

N ? ☑   N/A

EligibleAge ? ☑   18   +

PSRange ? ☑   70   100

EligibleNumberMets ? ☑   0   +

Recruit%HighKPS ? ☑   0   %

EligiblePS ? ☑   ● KPS ○ ECOG

Recruit%SingleMet ? ☑   0   %

70   100

Recruit%Male ? ☑   0   %

ExcludedHistology ? ☑   Germ-cell tumour
Germinoma
Leptomeningeal disease
Leukaemia

Recruit%NSCLung ? ☑   0   %

ExtraCranialDisease ? ☑   Absent
Diagnosed concurrently with brain involv
Stable one month
Stable two months

Recruit%Breast ? ☑   0   %

Recruit%Testicular ? ☑   0   %

EligibleRPA ? ☑   1   2

Recruit%Renal ? ☑   0   %

RecruitmentArea ? ☑

YearStarted ? ☑   2014

Recruit%StableECD ? ☑   0   %

YearStopped ? ☑   2014

Submit

URLToPaper ? ☑   N/A

Reset Form

Fig. 3 Adding a new clinical trial

4   Performing Criteria Searches

In the previous section, we have discussed how users can add entries into the system and edit existing ones. We will now focus on how we can leverage this information to the users' advantage. Initially, RecoMedic was supposed to exclusively be a decision support system for selecting medical literature, but following our discussions with experienced oncologists we realized that we could greatly improve the system by providing other useful features as well.

Our collaborators pointed out that while the clinical trials are easily accessible, finding specific information regarding them is not as trivial. Suppose a user wants to retrieve clinical trials with two arms for which the subjects are over sixty years old and where at least fifty percent of patients have a high Karnofsky Performance Status

(KPS)[7] score. Finding the appropriate literature concerning these trials in the old fashioned way would involve a substantial effort on the user's part and would mean browsing through numerous scientific papers. However, since our system already collects data concerning clinical trials, we can use our database to answer specific queries such as the one we have described.

The system presents the results retrieved for each query in tabular form directly in the web-based UI. Furthermore, the results can be exported into CSV format and downloaded for offline viewing and manipulation using popular tools such as Microsoft Excel.

5 Computing and Explaining Decisions

The central functionality of RecoMedic is represented by making and justifying patient-based medical literature recommendations. The fundamental condition for doctors to embrace a new technology or innovation is to first have it earn their trust. Our web platform strives to be transparent by not only recommending the most relevant pieces of literature, but also by providing natural language explanations as to why they were selected. Our users learn how the system "thinks" and they can easily pinpoint and suggest ways to improve it when they disagree with the decisions it recommends.

For our purposes, we decided to use the notion of *most preferred decisions* that was introduced in [7] (see Sect. 2), so that our system can reason about patient characteristics according to user-specified preferences on their characteristics.

The possible decisions in our framework ($d_k \in D$) are represented by the medical literature. The goals ($g_j \in G$) are the inputs in our system given by the individual patient characteristics that we would like for the literature to reflect. The decisions may have some attributes ($a_i \in A$) and each goal is satisfied by some attributes according to the tables (T_{DA} and T_{GA}) that we introduced in Sect. 2. For the current version of our system we focus on four patient characteristics (goals): age, primary disease, number of metastases and performance status score. After consulting with oncologists, we decided to use them as follows:

- The goal on age can be satisfied by two possible attributes: over eighteen years old (adult) and under eighteen years old (minor) which are extracted directly from the *EligibleAge* parameter of each clinical trial.
- The goal on primary disease can be satisfied by four possible attributes: breast, lung, testicular and renal cancer which are deduced from the corresponding recruit percentages in the clinical trial arms depending on whether these pass a certain threshold (we chose 60 %).

---

[7]This is one of the measures of patients' performances, and can range from 100 meaning 'normal' to 0 meaning 'death'.

- The goal on number of metastases can be satisfied by four possible attributes: no mets, one met, two mets, over two mets which are extracted from the *EligibleNumberMets* parameter of clinical trials.
- The goal on performance status score can be satisfied by four possible attributes: ECOG PS zero or one, ECOG PS two, ECOG PS three or four extracted from the *EligiblePS* parameter of clinical trials. The performance status scores expressed in KPS scale have been converted to their ECOG equivalent.[8]

We have created two user classes for RecoMedic. The first category is represented by healthcare specialists who need not concern themselves with the specifics of ABA. Indeed, we use ABA for recommending the appropriate literature, but we display the recommendation and explanation in natural language as can be seen in Fig. 4. The natural language explanation is created by examining the output produced by our argumentative computation engine corroborated with facts extracted directly from the generated ABA framework. It presents the patient characteristics according to the user specified ranking and states whether the recommended medical literature matches those certain patient characteristics. The phrasing includes both the user selection and the clinical trial retrieved value for that respective attribute, in a manner that we believe is clear and concise.

Our second category of users, computer scientists, can dive into the intrinsics of ABA by viewing the exact ABA framework that is generated for each query, they can view the output produced by our argumentative engine (sanctioning an argument as admissible, see Sect. 2), and can even choose between two possible argumentative engines for ABA (*proxdd* and *grapharg* [10][9]). A sample debate graph that results from executing a query for the computer scientist user class is illustrated in Fig. 5. Here we selected a 55 year old patient with lung cancer with 3 metastases and performance status (ECOG) 3, with the preferences ranked as follows: primary disease, number of metastases, age and performance status. The debate graph shows how the proponent in favor of selecting paper 40 (Langley ClinOnc [11]) successfully defends against the attacks of the opponent suggesting other papers. For example, the opponent's argument that paper 40 does not address patients with 3 metastases, or that paper 18 (Aoyama Jama [12]) addresses a higher preference goal, is quickly refuted by the proponent who, based on the facts in the knowledge base, can state that paper 40 indeed addresses patients with 3 metastases. The bottom three arguments by the opponent in Fig. 4 are not explicitly attacked because they have been previously defeated by the proponent when attacking the top three arguments in the figure.

---

[8]A detailed comparison between ECOG and KPS scales is available at http://oncologypro.esmo.org/Guidelines-Practice/Practice-Tools/Performance-Scales.

[9]Both are available at http://www.doc.ic.ac.uk/~rac101/proarg/.

Trial 18 is the best match.

Your most important factor was the primary disease. Trial 18 focuses on Lung cancer patients, and your patient has Lung cancer.

Your second most important factor was the number of metastases. Trial 18 is about patients with more than two metastases, and your patient has 3 metastases.

Your third most important factor was the performance status. Trial 18 is aimed at patients with (ECOG equivalent) PS two, and your patient has score 2.

Your fourth most important factor was age. Trial 18 focuses on adults, and your patient is 71 years old.

Fig. 4 Recommending a clinical trial and justifying the decision (healthcare specialist view)

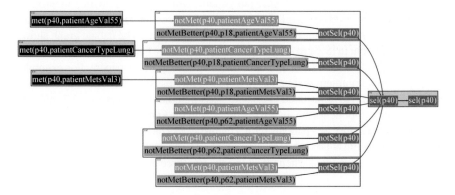

Fig. 5 Rendered ABA debate graph (computer scientist view)

6 Comparing Decisions

Besides recommending a certain piece of medical literature according to the characteristics of a given patient, the users may also want to assess how the suitability of
one clinical trial compares to another with respect to that patient. For this reason, we
decided to introduce a pairwise comparison feature illustrated in Fig. 6, that examines two pieces of literature at a time and provides a similarity score based on the
selected patient characteristics and the ranked user preferences.

Since we are using the notion of most preferred decision, meaning that satisfying
the higher ranked goal is more important than satisfying all the goals which are
ranked lower (see Sect. 2), we can score the decisions as follows:

$$score_d = \sum_{j=1}^{n} \#_{met(d,g_j)} \times 2^{n-rank(g_j)}$$

Decision/Justification   Pairwise Comparison

How does Trial 18 compare to    **?**   Explain

Comparison

Your most important factor was the primary disease. Trial 18 focuses on Lung cancer patients
while Trial 21 does not.

Your second most important factor was the number of metastases. Trial 18 is about patients with
more than two metastases while Trial 21 is not.

Your third most important factor was the performance status. Trial 18 is aimed at patients with
(ECOG equivalent) PS two and so is Trial 21.

Your fourth most important factor was age. Trial 18 focuses on adults while Trial 21 does not.

Trial 18 is more suitable than Trial 21 by score 13.

Score interpretation:

1-5 : quite similar

6-10 : superior to

11-15 : vastly superior to

Fig. 6 Comparing clinical trial suitability based on patients

Here, n is the number of ranked preferences and $rank(g_j)$ is an integer between 1 (if g_j is the top preference) and n (if g_j is the last ranked preference). This means that the score of clinical trial d (decision) is equal to the weighted sum of the goals that it meets, where $\#_{met(d,g_j)}$ is equal to 1 if g_j meets d. If a goal is not met by a decision, then $\#_{met(d,g_j)}$ equals 0 and that term is ignored having no effect on the clinical trial score.

We derive a simple similarity score between two decisions using the individual scores of each decision. Thus for two decisions d and d' the similarity score is:

$$similarity_{d,d'} = |score_{d'} - score_{d'}|$$

Using the similarity score we can assess how much alike or different the two given clinical trials are.

In our case, we are using four patient characteristics and the similarity score can be in the following ranges:

1–5;—the two clinical trials are similar
6–10;—one clinical trial is superior to the other
11–15.—one clinical trial is vastly superior to the other

7   System Design and Implementation

In Fig. 7 we present an overview of the system architecture. On the client side the users can fill in the patient details and receive the recommendation and justification once the computation is complete.

On the server side, we need multiple components to ensure that all the pieces needed to make decisions are present in the platform. First, we need to include a connection to a relational database management system so that clinical trial information can be easily transferred to and from the application. We have chosen to use an extra persistence layer between our server and the actual database in order to make the a seamless transition between objects and table rows. Next, we needed to create a custom grounder that was capable of mapping the ABA framework expressed as schemata (see Sect. 2 onto the grounded ABA framework, without schemata, that the argumentation engines manipulate. The grounder also acts as the entry point into the distributed platform. Once the grounded ABA framework is constructed we can then feed it into the argumentative engine. The engine can also display the output in graphical form as a debate graph (see Fig. 5), by outputting a DOT format debate graph which can in turn be viewed by our users. We chose to render the debate graph on the client side, in order to relieve the server the burden of additional processing. Finally, we have included a Natural Language Converter which produces an informal explanation as to why a certain decision was reached (see Figs. 4 and 6).

RecoMedic was designed to accommodate both the pressing oncologist needs by providing the means to add, edit or delete clinical trials, search for literature in

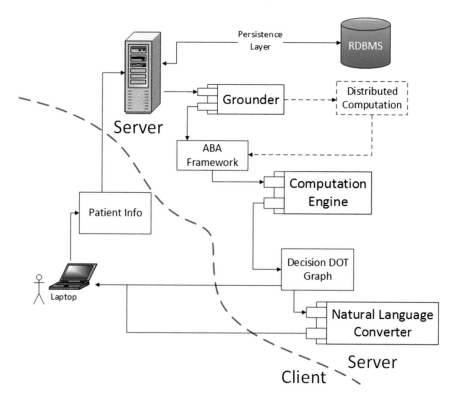

Fig. 7 General system architecture

the local repository, recommend medical literature based on individual patients and compare literature based on these patients, but at the same time it enables us to develop experimental features such as distributed ABA dialogue decision-making discussed in [8]. In the multi-agent decision making approach, the agents involved (e.g. different doctors or hospitals) deal with incomplete or imperfect information by exchanging messages to produce a joint ABA framework. Any conflicts that arise in the individual knowledge bases are dealt with by using a hierarchical trust scheme, in which one agent is considered more trustworthy than its counterpart. The entry point for the distributed computation is illustrated in Fig. 7 by the dotted line and the result is a jointly-produced, grounded ABA framework.

In Fig. 8 we have outlined the main technological components that were used in the development of RecoMedic. On the client side we have chosen HTML/CSS and Twitter's Bootstrap framework along with scripting in JavaScript, JQuery and AJAX for creating an intuitive, interactive and responsive web interface. Since our platform is mostly Java based, on the server side we use Jasypt for encrypting sensitive information such as user passwords. Moreover, Apache Maven enables automatic dependency resolution, making RecoMedic easily deployable on a different server. Also contributing to this, Hibernate makes the transition from the object oriented

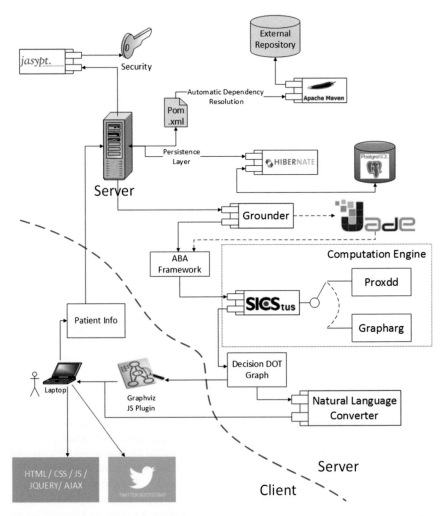

Fig. 8 Technologies used in system architecture

application to the relational database, represented by the open source PostgreSQL in our case. However, using Hibernate as our Object Relational Mapping (ORM) framework we can easily switch to another database vendor by changing one line in the configuration file. For making the recommendations we use two internal computation engines, *proxdd* and *grapharg* which produce the decision DOT graphs. These are interpreted by our custom Natural Language Converter on the server side or sent to the open source *Viz.js*[10] GraphViz JavaScript plugin for rendering on the client side.

---

[10]Viz.js is available at https://github.com/mdaines/viz.js/.

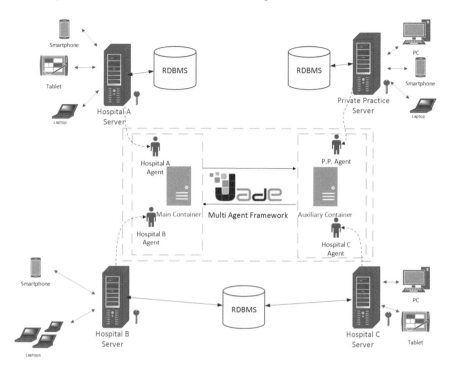

Fig. 9 Distributed system architecture

The Jade multi-agent framework [13][11] allows us to support distributed computation. It has built-in features such as agent discovery, complex agent behaviours and interaction protocols which make it an ideal choice for our system. We envisioned our platform to run autonomously in different healthcare institutions which can be interconnected through the multi-agent system. Any interaction with the outside world will be done through the representative agent of the institution, which could be replaced by a pool of agents should a single agent prove to be a bottleneck.

In Fig. 9 we have depicted our chosen configuration for deploying the distributed computation. In this diagram we can observe that some institutions may share their database with others, and that JADE agent containers may host the agents of multiple institutions on criteria such as geographic proximity or medical specialization. Although functional, the distributed extension of our system was not included in our user evaluation as more participating institutions are needed in order to produce meaningful results.

---

[11] Jade is available at http://jade.tilab.com/.

| Table 1 | 11 medical studies on brain metastases |

Id	ArmID	PMID number
1	Aoyama Jama 2006 [12]	16757720
2	Graham IJROBP 2010 [14]	19836153
3	Chang Lancet 2009 [15]	19801201
4	Langley ClinOnc 2013 [11]	23211715
5	Kocher JCO 2011 [16]	21041710
6	Patchell NEJM 1990 [17]	2405271
7	Patchell Jama [18]	9809728
8	Mintz Cancer 1996 [19]	8839553
9	VechtAnn Neurol 1993 [20]	8498838
10	Andrews Lancet 2004 [21]	15158627
11	Kondziolka IJROBP 1999 [22]	10487566

8 Evaluation

For the evaluation, we have identified 11 randomized clinical trials on the treatment of brain metastases. The decisions of our model are choices to use a given paper in a diagnosis—they can therefore be represented by names or IDs for the papers themselves. The Arm IDs and PMID Numbers of these papers from the literature are given in Table 1. Each paper contains a two-arm trial.

Based on these medical literature data, we have conducted a survey on user experience with our system. A small group of postgraduate medical students from Southern Medical University in China were invited to participate in the evaluation. An online survey system[12] was used for this evaluation.

The overall impression of the system is positive as 100 % of the users believe that RecoMedic serves a genuine need of doctors in searching for clinical trials for patients. 78.75 % users believe that they are able to fully understand the purpose and all of the functions of RecoMedic without any struggle. On a scale from 0 to 10, the users rank the UI design of My Clinical Trail, Recommend Trails, Criteria Search and Add Clinical Trial 7.63, 7.44, 7.4 and 7.1, respectively. Specifically, none of the users has considered that the overall UI design is complicated.

9 Related Work

Our work sits at the intersection between existing work on modeling clinical trials, and logic-based approaches to reasoning with clinical knowledge, specifically clinical trials. Much of the existing work on modeling clinical trials has been based

---

[12] Available at http://www.sojump.com.

around defining inclusion criteria to allow for (semi-)automated matching of patients with clinical trials [23],[13] and has largely been based on ontological approaches. Their work is based on whether a patient meets absolute inclusion/ exclusion criteria, whereas ours is based on how well a patient matches different patient populations. However, we note that the encoding of trial data is compatible with suggested approaches, such as the "Human Studyome" project,[14] although our approach to reasoning goes beyond their aims of knowledge encoding. The BioBIKE project[15] aims to capture and reason with biomedical data, but their focus is on biological and genomic data rather than clinical trials.

Other work that has focused on clinically-orientated reasoning includes many systems that provide medical knowledge representation and reasoning. However, many of these systems require specialist encoding of medical knowledge (e.g. [24, 25]) and hence divorce the clinicians from the underlying clinically-produced knowledge. The work closest to ours [26, 27] also works with encoded clinical trial data, but focuses on reasoning with the trial results in order to decide on the best treatment. The work we have presented here, which focuses on selecting the most appropriate paper, is orthogonal to this, and could be used as an input to their work, providing preferences (in this case based on "paper appropriateness") over conflicting trials.

10 Conclusions

In this paper, we have presented an online medical literature recommendation system, RecoMedic, tailored to patients with brain metastases. Through its easy to use web interface, RecoMedic allows users, primarily doctors, to search through its medical literature repository with patient information. This search is realized with argumentation such that not only the most relevant medical article is identified, but also an explanation to this selection is provided. Standard data repository maintenance features are also supported in RecoMedic, including adding, searching through, and comparing medical papers.

The implementation features a back end that interfaces with a Prolog based computation engine, and which runs JADE agents as part of a distributed solution, while the web-based front end is designed to provide our users with easy access to all the system features.

A preliminary user evaluation has confirmed that RecoMedic addresses a genuine need for doctors. The evaluation also confirms that users are able to use RecoMedic with little help and approve its UI design.

In the future, we would like to experiment RecoMedic with more users and extend its features to include not only recommending most relevant articles, but also most suitable treatments. We would also like to explore other search and ranking strategies

---

[13] Available at http://bioportal.bioontology.org/ontologies/OCRE.

[14] Available at http://rctbank.ucsf.edu/.

[15] Available at http://biobike.csbc.vcu.edu/.

and explore the use of argumentation in these strategies. We also would like to explore our approach to medical conditions other than brain metastases.

Acknowledgments The first author was supported by the strategic grant POSDRU/159/1.5/S/ 133255, Project ID 133255 (2014), co-financed by the European Social Fund within the Sectorial Operational Program Human Resources Development 2007–2013.
The second and third authors were supported by the EPSRC project *Transparent Rational Decisions by Argumentation*: EP/J020915/1.

References

1. Yildirim, P., Majnaric, L., Ekmekci, O., Holzinger, A.: Knowledge discovery of drug data on the example of adverse reaction prediction. BMC Bioinform. **15**(Suppl 6), S7 (2014)
2. Bench-Capon, T.J.M., Dunne, P.E.: Argumentation in artificial intelligence. AIJ **171**(10–15), 619–641 (2007)
3. Besnard, P., Hunter, A.: Elements of Argumentation. The MIT Press (2008)
4. Rahwan, I., Simari, G.R.: Argumentation in Artificial Intelligence. Springer (2009)
5. Dung, P.M., Kowalski, R.A., Toni, F.: Assumption-based argumentation. In: Argumentation in AI, pp. 199–218. Springer (2009)
6. Fan, X., Toni, F.: Decision making with assumption-based argumentation. In: Proceedings of the TAFA (2013)
7. Fan, X., Craven, R., Singer, R., Toni, F., Williams, M.: Assumption-based argumentation for decision-making with preferences: a medical case study. In: Proceedings of the CLIMA (2013)
8. Fan, X., Toni, F., Mocanu, A., Williams, M.: Multi-agent decision making with assumption-based argumentation. In: Proceedings of the AAMAS (2014)
9. Matt, P.A., Toni, F., Vaccari, J.: Dominant decisions by argumentation agents. In: Proceedings of the ArgMAS. Springer (2009)
10. Craven, R., Toni, F., Williams, M.: Graph-based dispute derivations in assumption-based argumentation. In: Proceedings of the TAFA (2013)
11. Langley, R.E., Stephens, R.J., Nankivell, M., Pugh, C., Moore, B., Navani, N., Wilson, P., Faivre-Finn, C., Barton, R., Parmar, M.K., Mulvenna, P.M.: QUARTZ Investigators. Interim data from the Medical Research Council QUARTZ Trial: does whole brain radiotherapy affect the survival and quality of life of patients with brain metastases from non-small cell lung cancer? Clin. Oncol. (R Coll Radiol). **25**(3), e23-30 (2013). doi:10.1016/j.clon.2012.11.002. Epub 2012 Dec 2. PubMed PMID: 23211715
12. Aoyama, H., Shirato, H., Tago, M., Nakagawa, K., Toyoda, T., Hatano, K., Kenjyo, M., Oya, M., Hirota, S., Shioura, H., Kunieda, E., Inomata, T., Hayakawa, K., Katoh, N., Kobashi, G.: Stereotactic radiosurgery plus whole-brain radiation therapy vs stereotactic radiosurgery alone for treatment of brain metastases: a randomized controlled trial. JAMA **295**(21), 2483–2491 (2006). PubMed PMID: 16757720
13. Bellifemine, F., Caire, G., Greenwood, D.: Developing Multi-Agent Systems with JADE (Wiley Series in Agent Technology). Wiley (2007)
14. Graham, P.H., Bucci, J., Browne, L.: Randomized comparison of whole brain radiotherapy, 20 Gy in four daily fractions versus 40 Gy in 20 twice-daily fractions, for brain metastases. Int. J. Radiat. Oncol. Biol. Phys. **77**(3), 648–654 (2010). doi:10.1016/j.ijrobp.2009.05.032. Epub 2009 Oct 14. PubMed PMID: 19836153
15. Chang, E.L., Wefel, J.S., Hess, K.R., Allen, P.K.: PK, F.F. Lang, D.G. Kornguth, R.B. Arbuckle, J.M. Swint, A.S. Shiu, M.H. Maor, C.A. Meyers. Neurocognition in patients with brain metastases treated with radiosurgery or radiosurgery plus whole-brain irradiation: a randomised controlled trial. Lancet Oncol. **10**(11), 1037–1044 (2009). doi:10.1016/S1470-2045(09)70263-3. Epub 2009 Oct 2. PubMed PMID: 19801201

16. Kocher, M., Soffietti, R., Abacioglu, U., Vill, S., Fauchon, F., Baumert, B.G., Fariselli, L., Tzuk-Shina, T., Kortmann, R.D., Carrie, C., Ben Hassel, M., Kouri, M., Valeinis, E., van den Berge, D., Collette, S., Collette, L., Mueller, R.P.: Adjuvant whole-brain radiotherapy versus observation after radiosurgery or surgical resection of one to three cerebral metastases: results of the EORTC 22952-26001 study. J. Clin. Oncol. **29**(2), 134-41 (2011). doi:10.1200/JCO.2010.30.1655. Epub 2010 Nov 1. PubMed PMID: 21041710; PubMed Central PMCID: PMC3058272

17. Patchell, R.A., Tibbs, P.A., Walsh, J.W., Dempsey, R.J., Maruyama, Y., Kryscio, R.J., Markesbery, W.R., Macdonald, J.S., Young, B.: A randomized trial of surgery in the treatment of single metastases to the brain. N. Engl. J. Med. **322**(8), 494–500 (1990). PubMed PMID: 2405271

18. Patchell, R.A., Tibbs, P.A., Regine, W.F., Dempsey, R.J., Mohiuddin, M., Kryscio, R.J., Markesbery, W.R., Foon, K.A., Young, B.: Postoperative radiotherapy in the treatment of single metastases to the brain: a randomized trial. JAMA **280**(17), 1485–1489 (1998). PubMed PMID: 9809728

19. Mintz, A.H., Kestle, J., Rathbone, M.P., Gaspar, L., Hugenholtz, H., Fisher, B., Duncan, G., Skingley, P., Foster, G., Levine, M.: A randomized trial to assess the efficacy of surgery in addition to radiotherapy in patients with a single cerebral metastasis. Cancer **78**(7), 1470–1476 (1996). PubMed PMID: 8839553

20. Vecht, C.J., Haaxma-Reiche, H., Noordijk, E.M., Padberg, G.W., Voormolen, J.H., Hoekstra, F.H., Tans, J.T., Lambooij, N., Metsaars, J.A., Wattendorff, A.R., et al.: Treatment of single brain metastasis: radiotherapy alone or combined with neurosurgery? Ann. Neurol. **33**(6), 583–590 (1993). PubMed PMID: 8498838

21. Andrews, D.W., Scott, C.B., Sperduto, P.W., Flanders, A.E., Gaspar, L.E., Schell, M.C., Werner-Wasik, M., Demas, W., Ryu, J., Bahary, J.P., Souhami, L., Rotman, M., Mehta, M.P., Curran Jr, W.J.: Whole brain radiation therapy with or without stereotactic radiosurgery boost for patients with one to three brain metastases: phase III results of the RTOG 9508 randomised trial. Lancet **363**(9422), 1665–1672 (2004). PubMed PMID: 15158627

22. Kondziolka, D., Patel, A., Lunsford, L.D., Kassam, A., Flickinger, J.C.: Stereotactic radiosurgery plus whole brain radiotherapy versus radiotherapy alone for patients with multiple brain metastases. Int. J. Radiat. Oncol. Biol. Phys. **45**(2), 427–434 (1999). PubMed PMID: 10487566

23. Tu, S.W., Carini, S., Rector, A., Maccallum, P., Toujilov, I., Harris, S., Sim, I.: Ocre: an ontology of clinical research. In: 11th International Protege Conference (2009)

24. Buchanan, B.G., Shortliffe, E.H.: Rule Based Expert Systems: The Mycin Experiments of the Stanford Heuristic Programming Project. Addison-Wesley Longman Publishing Co., Inc., Boston (1984)

25. Grando, M.A., Glasspool, D., Fox, J.: A formal approach to the analysis of clinical computer-interpretable guideline modeling languages. AIM **54**(1), 1–13 (2012)

26. Hunter, A., Williams, M.: Aggregating evidence about the positive and negative effects of treatments. AIM **56**(3), 173–190 (2012)

27. Gorogiannis, N., Hunter, A., Williams, M.: An argument-based approach to reasoning with clinical knowledge. IJAR **51**(1), 1–22 (2009)

Medical Image Processing: A Brief Survey and a New Theoretical Hybrid ACO Model

Camelia-M. Pintea and Cristina Ticala

Abstract The current paper includes a brief survey on image processing, in particular for medical image processing, including the main algorithms on segmentation and margin detection. Both mathematical background and algorithms are detailed. Some of the most efficient ant-based algorithms used for image processing are also described. It is also introduced a new theoretical hybrid Ant Colony Optimization model in order to enhance medical image processing. The newly introduced model uses artificial ants with different levels of "sensitivity" and also a model of "direct" communication as in Multi-Agent Systems.

1 Introduction

Many researchers are working today in *Image Processing*. It is used in a large area of real-life domains as industry, economy, medicine etc. In particular, *image processing* in medicine is used frequently (e.g. tomography) and it is intended to solve patients medical problems.

Tomography refers to imaging by sections or sectioning, by the use of any kind of penetrating wave. A device used in tomography is called a tomograph. The image produced by tomograph is called tomogram. The computed tomographic (CT) scanner was invented by Sir Godfrey Hounsfield and it is an exceptional contribution to medicine [3]. Tomography is used in radiology [27], biology [37], geophysics [26], plasma physics [50], archeology [24], materials science [38], oceanography [29], astrophysics [50], quantum information [30], and other sciences. It is based on the mathematical procedure called tomographic reconstruction.

C.-M. Pintea (✉) · C. Ticala
Faculty of Sciences, TU Cluj-Napoca, Cluj-napoca, Romania
e-mail: dr.camelia.pintea@ieee.org

C. Ticala
e-mail: cristina.ticala.pop@gmail.com

© Springer International Publishing Switzerland 2016
I. Hatzilygeroudis et al. (eds.), *Combinations of Intelligent Methods and Applications*, Smart Innovation, Systems and Technologies 46,
DOI 10.1007/978-3-319-26860-6_7

117

Images iterative reconstruction refers to iterative algorithms which are used to reconstruct 2D and 3D images in certain imaging techniques. For example, in computed tomography an image must be reconstructed from projections of an object. Iterative reconstruction techniques provide better images than the common filtered back projection (FBP) method. Iterative reconstruction algorithms are computationally are more expensive than FBP which directly calculates the image in a single reconstruction step.

Coincidence events may be grouped into projection images [17], called sinograms. The sinograms are sorted by the angle of each view. The sinogram images are similar to the projections captured by computed tomography (CT) scanners, and can be reconstructed in a resembling way.

A normal PET (Positron Emission Tomography) data set has millions of counts for the whole acquisition, while the CT can reach a billion counts. This contributes to PET images appearing "noisier" than CT. Scatter and random events are two major sources of noise in PET. Scatter is generated when from a detected pair of photons, at least one was deflected from its original path by interaction with matter in the field of view, leading to the pair being assigned to an incorrect LOR. Random events appear when photons originating from two different annihilation events are incorrectly recorded as a coincidence pair because their arrival at their respective detectors occurred within a coincidence timing window.

The current paper illustrates the main algorithms used for image processing including the algorithms inspired by ant colonies. It is also included a new ant algorithm with both sensitive features and "direct" communication (from Multi-agent System). The hybrid ant-based model will use the *Sensitive Ant Model* [10] successfully used for solving several complex problems as for example Denial Jamming Attack on Wireless Sensor Network [44], Sensor networks security [40]. There are several variants of sensitive ant models as for example *Sensitive Stigmergic Agent Systems* [41], *Cooperative Learning Sensitive Agent System for Combinatorial Optimization* [12, 41] and *a Step-Back sensitive ant model* [9, 41].

This paper has the following structure. The next section is about the state-of-art in medical image processing, with the most efficient existing algorithms; Sect. 3 is about the main ant-based models used today to solve the problems of segmentation and margin detection of medical images. In Sect. 4, it is introduced a newly theoretical model of Ant Colony Optimization, including several agents features, from Multi-Agent System and some particular features as sensitivity.

2   State-of-Art in Image Processing

This section starts with a short introduction of medical images, as tomography, and the main algorithms used to medical image processing including tomographic images.

Tomography, for example a radiography (Fig. 1), supposes projection data acquisition from multiple directions and feeding the data into a software of tomographic reconstruction processed by a computer. When there are used X-rays, the tomogram

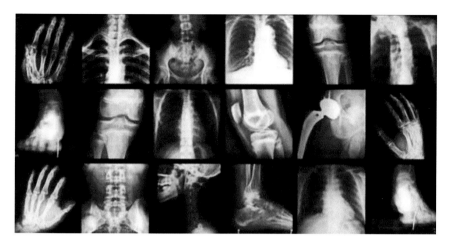

Fig. 1 Examples of images from radiology [28]

is called X-ray computed tomogram (CT); single-photon emission computed tomography (SPECT) is a tomographic imaging technique which uses gamma-rays. Magnetic resonance imaging (MRI) scanners use strong magnetic fields and radio waves in order to obtain images of the object. Positron emission tomography (PET) is a functional imaging technique in nuclear medicine that produces a three-dimensional image of functional processes in the body. Electrical resistivity tomography (ERT) or electrical resistivity imaging (ERI) is a geophysical technique used for sub-surface imaging by making electrical resistivity measurements on the ground surface [25] (Fig. 2).

X-ray CT is a technique which virtually slices the scanned object in order to produce images of some specific areas (Fig. 3). The rays are computer processed and they allow the user to see inside the object without cutting it. From a large series of two—dimensional images taken around an axis of rotation there are generated three—dimensional images of the inside of the object by using digital geometry processing. The most common application of CT is medical imaging. The images are used for diagnostic in various medical disciplines [22].

In 1917, the Austrian mathematician Johann Radon invented the Radon Transform [47, 48]. He proved mathematically that a function could be reconstructed from a finite set of its projections [23]. In 1937 a Polish mathematician, Stefan Kaczmarz, developed a method to find an approximate solution for a large system of algebraic equations [32, 33]. Due to the mathematical theory related to Radon Transform and to Kaczmarz algorithm, computed tomographic image reconstruction was possible. Kaczmarz algorithm led to another powerful reconstruction method called "Algebraic Reconstruction Technique (ART)" which was adapted by Sir Godfrey Hounsfield as the image reconstruction method in the first commercial CT scanner.

For image reconstruction the positron emission tomograph (PET) we need a list of 'coincidence events'. These are the data collected and they represent almost simulta-

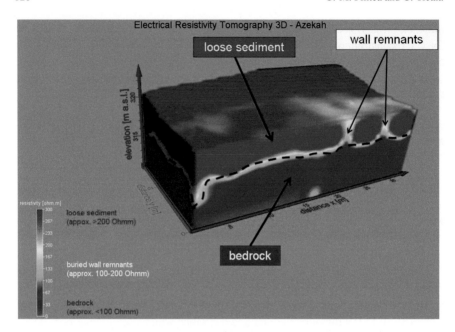

Fig. 2 Geophysical prospection by means of electrical resistivity tomography [24]

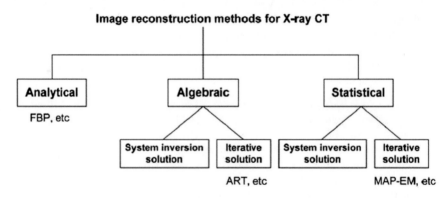

Fig. 3 Image reconstruction methods for X-ray CT [22]

neous detection of photons by a pair of detectors. Each coincidence event represents a line in space connecting the paired detectors.

About Filtered Back Projection (FBP)

Filtered back projection (FBP) has been often used to reconstruct images from the projections. *FPB* basic idea is to simply run the projections back through the image to obtain a rough approximation to the original. The projections will interact constructively in regions that correspond to the emitting sources in the original image. A problem is the star-like artifacts which occur in the reconstructed image, where the back projections intersect. In practice, the collected projections cannot be used for CT reconstruction directly without pre-processing [54]. Correction for random coincidences, estimation and subtraction of scattered photons, detector dead-time correction (after the detection of a photon, the detector must "cool down") and detector-sensitivity correction must be applied [7]. Its advantage is simplicity and low requirement for computing resources. However, in practice, noise is prominent in the reconstructed images. Also, FBP treats the data deterministic—it does not account for the inherent randomness associated with PET data, thus requiring all the pre-reconstruction corrections.

Iterative expectation-maximization algorithms are now the popular method of reconstruction. These algorithms compute an estimate of the distribution of anni-hilation events that led to the measured data, based on statistical principles. The advantage of iterative algorithms is a better noise profile than FBP, but the disadvan-tage is higher computer resource requirements [51].

Attenuation occurs when photons emitted by the tracer inside the body are absorbed by intervening tissue between the detector and photon emission. As different LORs must traverse different thicknesses of tissue, the photons are attenuated differentially. As a result, structures deeper in the body are reconstructed as having falsely low tracer uptake. While attenuation-corrected images are generally more exact representations, the correction process is itself susceptible to significant errors. As a result, both corrected and uncorrected images are always reconstructed and read together.

Early PET scanners had only a single ring of detectors, hence the acquisition of data and the reconstruction was restricted to a single transverse plane. More modern scanners now include multiple rings, essentially forming a cylinder of detectors. There are two approaches to reconstructing data from such a scanner:

(1) each ring is treated as a separate entity, only coincidences within a ring are detected and the image from each ring can then be reconstructed individually ($2D$ reconstruction), or
(2) coincidences will be detected between rings as well as within rings, then recon-struction will represent the entire volume together ($3D$).

The $3D$ techniques are better because more coincidences are detected and used; as disadvantages are their sensitivity to the effects of scatter, random coincidences and large computational resources. This image reconstruction is encouraged by the advent of the particle detectors with sub-nanosecond time resolution which afford better random coincidence rejection.

In the mathematical description of the image reconstruction problem in transmis-sion tomography a popular approach is the algebraic formulation of the problem [7]. Here the problem is to solve a large system of linear equations, $Ax = b$.

Let see the body in $2D$ as an image with finitely number of squares or pixels and in $3D$ formulation as consisting of finitely many cubes, or voxels [5]. In $2D$ the attenuation function is discretized, in the two-dimensional case where the function has an unknown constant value denoted x_j at the jth pixel. In $3D$ [5] the beam is sent through the body along various lines and both initial and final beam strength is measured. Further is calculated a discrete line integral along each line. Let denote L_i the ith line segment through the body and by b_i its associated line integral (Eq. (1)); A_{ij} is the length of the intersection of the jth pixel with L_i; A_{ij} is nonnegative. Most pixels do not intersect line L_i, so A is quite sparse. Both I, the number of lines and the number of pixels J are large and typically unrelated.

$$b_i = \sum_{j=1}^{J} A_{ij} x_j \tag{1}$$

The matrix A is large, rectangular and the system $Ax = b$ may or may not have exact solutions. The number of pixels, J, could be large limited only by computation costs. The number I of lines are based on the constraints posed by the scanning machine and the desired scan duration and dosage.

For an underdetermined system ($J > I$) are infinitely many exact solutions, so some constraints and prior knowledge are used to select an appropriate solution. Noise in the data, error in the model of the physics of the scanning procedure gives undesirable solutions [4]. For an overdetermined system ($J < I$) are seeking some approximate solution. The physics of the materials present in the body provide upper bounds for x_j and information on body shape and structure to find where x_j is zero. Incorporating this information in the reconstruction algorithms can often lead to improved images [39].

In *Single-Photon Emission Tomography (SPECT)* and *Positron Emission Tomography (PET)* the patient is injected or inhales a chemical to which a radioactive substance has been attached. The recent book edited by Wernick and Aarsvold [53] describes the cutting edge of emission tomography. The particular chemicals used in emission tomography are designed to become concentrated in the particular region of the body under study. The radioactivity results in photons that travel through the body are detected by the scanner [1].

The function of interest is the actual concentration of the radioactive material at each spatial location within the region of interest. Tumors may take up the chemical and its radioactive passenger, more avidly than normal tissue, or less avidly. Malfunctioning brain portion may not receive the normal amount of the chemical and will exhibit an abnormal amount of radioactivity [5].

The nonnegative function is discretized and denoted as the vector x. The quantity b_i, the ith entry of the vector b, is the photon count at the ith detector; in coincidence detection *PET* a detection is a nearly simultaneous detection of a photon at two different detectors. The entry A_{ij} of the matrix A is the probability that a photon emitted at the jth pixel or voxel will be detected at the ith detector.

In [49], Rockmore and Macovski suggest that, in the emission tomography one take a statistical view, where x_j is the expected number of emissions at the jth pixel during the scanning time. The expected count at the ith detector is computed as in Eq. (2):

$$E(b_i) = \sum_{j=1}^{J} A_{ij} x_j. \tag{2}$$

The problem of finding the x_j may be viewed as a parameter-estimation problem for which a maximum-likelihood technique might be helpful [49]. These led to the *Expectation Maximization Maximum Likelihood (EMML)* method for reconstruction. The system of equations $Ax = b$ is obtained by replacing the expected count, $E(b_i)$, with the actual count, b_i. The system solution should be an approximate nonnegative one, with nonnegative entries.

The Original Kaczmarz Algorithm

In [46] consider the already mentioned system $Ax = b$. Let A be an $J \times J$ invertible matrix and b be a member of C^J. Let $x^* = b \cdot A^{-1}$ be the unique solution of the above system. We normalize its equations as follows in Eq. (3) by scaling a procedure as in Eq. (4).

$$\|A_i\|^2 = \sum_{m=1}^{J} A_{ij}^2 = 1, \tag{3}$$

$$D^{-1}Ax = D^{-1}b, \quad \text{with } D = diag\left(\|A_1\|, \ldots, \|A_J\|\right). \tag{4}$$

For x_0 as in Eq. (5) as an initial approximation we successively define $x^{(0,1)}, \ldots, x^{(0,J)} \in \mathbb{R}^J$ by Eq. (6).

$$x_0 = \left(x_1^{(0,0)}, x_2^{(0,0)}, \ldots x_J^{(0,0)}\right) \in \mathbb{R}^J \tag{5}$$

$$\begin{cases} x^{(0,1)} = x^{(0,0)} - \left[\langle x^{(0,0)}, A_1\rangle - b_1\right] A_1 \\ x^{(0,2)} = x^{(0,1)} - \left[\langle x^{(0,1)}, A_2\rangle - b_2\right] A_2 \\ \vdots \\ x^{(0,J)} = x^{(0,J-1)} - \left[\langle x^{(0,J-1)}, A_J\rangle - b_J\right] A_J \end{cases} \tag{6}$$

Then, for an arbitrary $r \geq 0$ and a given approximation $x^{(r,J)} \in \mathbb{R}^J$ are successively constructed the new ones $x^{(r+1,1)}, \ldots, x^{(r+1,J)} \in \mathbb{R}^J$ as in Eq. (7) with the convention $x^{r+1,0} = x^{r,J}$.

$$x^{(r+1,i)} = x^{(r+1,i-1)} - \left[\langle x^{(r+1,i-1)}, A_i\rangle - b_i\right] A_i, \quad \text{for all } i = 1, \ldots J. \tag{7}$$

Algebraic Reconstruction Technique

The problem is to solve a system of linear equations ($Ax = b$), with a simple method, the *Algebraic Reconstruction Technique (ART)*. The *ART* model was introduced by Gordon et al. [20] as a method for image reconstruction in transmission tomography and it is a special case of Kaczmarzs algorithm.

Let L_i be the set of pixel indices j for which the jth pixel intersects the ith line segment, and let $|L_i|$ be the cardinality of the set L_i. Let $A_{ij} = 1$ for j in L_i and

$A_{ij} = 0$ otherwise. With $i = k \, (mod \; I) + 1$, the iterative step of the ART algorithm is based on Eq. (8).

$$x_j^{k+1} = \begin{cases} x_j^k + \frac{1}{|L_i|} \left(b_i - \left(Ax^k \right)_i \right), \; j \in L_i \\ x_j^k, \; j \notin L_i \end{cases}, \tag{8}$$

where $k \, (mod \; I)$ is the remainder of the Euclidean division of k by I. In each step of ART, we take the error, $b_i - \left(Ax^k \right)_i$, associated with the current x_k and the ith equation, and distribute it equally over each pixels that intersects L_i.

The *ART* method can be viewed as an iterative method for solving an arbitrary system of linear equations, $Ax = b$. Let A be a complex matrix with I rows and J columns, and let b be a member of C^I. We want to solve the system $Ax = b$. For each index value i, let H_i be the hyperplane of J-dimensional vectors given by Eq. (9) and P_i the orthogonal projection operator onto H_i:

$$H_i = \{x | \, (Ax)_i = b_i\}. \tag{9}$$

Let x_0 be arbitrary; for each nonnegative integer k, let $i(k) = k \, (mod \; I) + 1$ the iterative step of the ART is Eq. (10):

$$x^{k+1} = P_{i(k)} x^k. \tag{10}$$

Given any vector z the vector in H_i closest to z, in the sense of the Euclidean distance, has the entries as in Eq. (11):

$$x_j = z_j + \overline{A_{ij}} \left(b_i - (Az)_i \right) / \sum_{m=1}^{J} |A_{im}|^2. \tag{11}$$

When the system $Ax = b$ has exact solutions the *ART* converges to the solution closest to x_0, in the 2-norm. How fast the algorithm converges will depend on the ordering of the equations. It is important to avoid particularly bad orderings where the hyperplanes H_i and H_{i+1} are nearly parallel [7].

Cimminos Algorithm

In *ART* it seeks a solution of $Ax = b$ by projecting the current vector x_k orthogonal onto the next hyperplane $H \left(a^{i(k)}, b_{i(k)} \right)$ to get x_{k+1}. In Cimminos algorithm is projected the current vector x_k onto each of the hyperplanes and then average the result to get x_{k+1} [18]. The algorithm starts with an arbitrary x_0; the iterative step is then Eq. (12) where P_i is the orthogonal projection onto $H(a^i, b_i)$:

$$x^{k+1} = \frac{1}{I} \sum_{i=1}^{I} P_i x^k. \tag{12}$$

The iterative step can then be written as in Eq. (13) where $A^\dagger$ is the conjugate transpose of the matrix A:

$$x^{k+1} = x^k + \frac{1}{l} A^\dagger \left(b - Ax^k \right). \tag{13}$$

One advantage of many simultaneous algorithms, such as Cimminos, is that they do converge to the least squares solution when the system $Ax = b$ does not have exact solutions [7].

The Landweber Algorithm

The Landweber algorithm [7] with the iterative step as in Eq. (14) converges to the least squares solution closest to the starting vector x^0. The provided value $0 < \gamma < 2/\lambda_{max}$ is the largest eigenvalue of the nonnegative-definite matrix $A^\dagger A$.

$$x^{k+1} = x^k + \gamma A^\dagger \left(b - Ax^k \right) \tag{14}$$

The EMML and SMART Algorithms

The *Expectation Maximization Maximum Likelihood (EMML), Simultaneous Multiplicative ART (SMART)* algorithms and *Rescaled Block—Iterative (RBI)* methods are based on the *Kullback-Leibler (KL)* distance between nonnegative vectors [7]. For $\alpha > 0$ and $\beta > 0$ the cross-entropy or *Kullback-Leibler* distance from α to β is as in Eq. (15) where $KL(\alpha, 0) = +\infty$, and $KL(0, \beta) = \beta$.

$$KL\left(\alpha, \beta \right) = \alpha \log \frac{\alpha}{\beta} + \beta - \alpha. \tag{15}$$

Let extend to nonnegative vectors coordinate-wise so that Eq. (16) stands.

$$KL(x, z) = \sum_{j=1}^{J} KL(x_j, z_j). \tag{16}$$

Unlike the *Euclidean distance*, the *KL distance* is not symmetric. Based on that different approximate solutions of $Ax = b$ are to be found by minimizing the two distinct distances, $KL(Ax; b)$ and $KL(b; Ax)$ with respect to nonnegative x.

The *EMML* algorithm minimizes $KL(b, Ax)$, while the *(SMART)* minimizes $KL(Ax, b)$ [6]. These methods were developed for application to tomographic image reconstruction, although they have much more general uses. Whenever there are nonnegative solutions of $Ax = b$ *SMART* converges to the nonnegative solution that minimizes $KL(x, x_0)$; *EMML* also converges to a non-negative solution, but no explicit description of that solution is known.

3   Ant-Based Models for Medical Image Processing

The field of "ant algorithms" studies models derived from the observation of real ants' behavior, and uses these models as a source of inspiration for the design of novel algorithms for the solution of optimization and distributed control problems.

Ant Colony Optimization (ACO) [16] is nowadays the most successful metaheuristic based on ant behaviors.

When an image contains many clustered objects, their overlapping can hide its structure. The existing segmentation techniques are not able to address implicitly the image parts, such that in [34] is proposed a hybrid approach called *Ant Colony Optimization and Fuzzy Logic* based technique.

Fuzzy Logic (FL) techniques have been used in image processing applications like edges detection, feature extraction, classification, and clustering. Fuzzy logic mimics the human mind to effectively employ modes of reasoning that are approximate rather than exact. A basic concept in *FL* is the fuzzy if-then rule [34]. *FL* can model nonlinear functions of arbitrary complexity to a desired degree of accuracy. See also another fuzzy approach with ACO [13, 14].

Combining the *Ant Colony Optimization* with *Fuzzy Logic* yields the structural information of the image implicitly. Using the *FL* a rule base is formed. When *ACO* is applied to the image, the autonomous agents collect each pixel intensity value. This pixel is assigned to a particular group based on the fuzzy rule. The complexity of this approach is forming the fuzzy rules. An algorithm explains the way how the grouping of the pixels is performed [34].

There are four rules dealing with the vertical and horizontal direction lines gray level values around the checked or centered pixel of the mask. From the side of the fuzzy construction, the input is ranged from 0 to 255 gray intensity, and according to the desired rules the gray level is converted to the values of the membership functions. The method [34] is implicitly removing the misclassifications and thus producing better results than the original *ACO* algorithm and it is extracting the features from the images implicitly. The advantages of this approach are: it has a very reduced set of fuzzy inference rule; avoids the mis-classifications of the intensities belonging to the overlapping regions and it is not required to use mask filter before processing.

In [8] it is proposed *Channeler Ant Model (CAM)* based on the natural ant capabilities of dealing with *3D* environments through self-organization and emergent behavior. It is already in use for the automated detection of nodules in lung *Computed Tomography*. The main advantage of the model is that provides an elegant solution for the segmentation of *3D* structures in noisy environments with unknown range of image intensities. The task of a *CAM* colony is to provide *3D* pheromone maps of the explored volume, to be used as a starting point for the structures segmentation. The *CAM* performance is better when the noise is lower and proves to be suitable for a full objects segmentation of different shape, intensity range in a noisy background [8].

Applications of the *ACO* to solve image processing problem with a reference to a new automatic enhancement technique based on real-coded particle ant colony is proposed in [21]. The optimization problem is to solve the enhancement problem using *ACO* and the objective is to maximize the number of pixels in the edges, increase the overall intensity of the edges, and increase the measure of the entropy. The obtained results indicate that the proposed *ACO* yields better results in terms of both the maximization of the number of pixels in the edges and the adopted objective evaluation when compare with *Genetic Algorithms (GAs)* and *Particle*

Swarm Optimization (PSO). *ACO* is more attractive in the way that there are few parameters to adjust compared with the large number of parameters adjusted when *PSO* and *GA* is run. The proposed *ACO* method yields high quality solutions with computation efficiency.

In [19] it is presented an extended model of an artificial ant colony system designed to evolve on digital image habitats. The swarm can adapt the population size according to the type of image on which it is evolving and it is reacting faster to changing images. The variation of the population size is achieved adapting the process of aging (and dying) as well as the reproduction process as it follows: a fixed amount of energy $e(0)$ is assigned to each ant created. Every generation, the energy $e(a)$ is updated, where a is the age of the ant measured in time steps. The value of $e(a)$ is computed by subtracting a fixed amount to the energy of the ant in the previous iteration, and by adding a dynamic value inspired in the pheromone trail. In the paper [19] there are examined the differences between Swarms with Fixed Population Size (SFPS) and Swarms with Varying Population Size (SVPS) when evolving on static environments.

In fewer generations, the maps of SVPS show a configuration that resembles the main lines of the original image figure [19]; for some images, the pheromone maps have different characteristics: the homogeneous regions have less noise in *SVPS* maps, giving the idea that ants are converging more efficiently to heterogeneous parts of the image. *SVPS* is more able to evolve sharp and noiseless pheromone maps that somewhat reflect the image contours. The varying population model is fast and more effective in creating pheromone trails around the edges of the images. The computational cost of the reproduction process is not significant and the running times *SFPS/SVPS* are very similar.

Edge detection is finding the points where there are sudden changes in the intensity values and linking them suitably. The paper [2] aims at presenting a comparison of the Gradient based existing edge detectors, with a swarm intelligence Ant Colony. The authors propose a new edge detector based on swarm intelligence, which fairly detects the edges of all types of images with improved quality, and with a low failing probability in detecting edges. The simple threshold technique is used here to partition the image histogram by a single global threshold T, segmentation is then accomplished by scanning the image, pixel by pixel and labeling each pixel as edge point or not depending on whether the gray level of that pixel is greater or less than the value of T.

Edge detecting in an image significantly reduces the amount of data and filters out useless information while presenting the important structural properties in an image. Edge detection is difficult in noisy images since both the noise and the edges contain high frequency content. Better results can be obtained by applying a noise filter prior to the edge detection [2].

Thresholding is the first step for many industrial problems. In [35] the authors propose an optimization algorithm combining parametric, *EM*—expectation-maximization and nonparametric approaches, *ACS-Otsu*—*Ant Colony System*. The *hybrid* [35] algorithm may be considered a two-phase bi-objective approach by embedding an *EM* algorithm into *ACS-Otsu* algorithm. At beginning *ACS-Otsu* searches the thresholds, level by level, for the given image based on between-class variance. When the predefined number of segmented levels is reached, the second phase starts and the new ants constructed so far will be evaluated based on two criteria.

At first each ant is evaluated by the between-class variance; secondly the initial parameters of each new ant will be evaluated by the curve fitting error and the best parameters are fed to the *EM* algorithm. The *EM* algorithm runs until its convergence criterion is reached. Once the *EM* algorithm stops the optimal thresholds are determinate. The second phase will continue until the maximum number of iterations in *hybrid algorithm* is reached. The *hybrid* algorithm is capable of providing quality segmentation within a stable and reasonable *CPU* time.

4   A New Theoretic Hybrid Ant-Based Sensitive Approach for Image Processing

The current section introduces a theoretical hybrid Ant Colony Optimization (ACO) [16] model to be used on medical image processing, called *Hybrid Medical Image Processing-Sensitive Ant Model* (*HMIP-SAM*). From the meta-heuristic *ACO* it is used in particular the *Ant Colony System (ACS)* heuristic. It involves a particular direct communication inspired from Multi-agent Systems and there are also used different levels of sensitivity of artificial ants, to improve the image processing solution.

4.1   Prerequisites

A short description of ACO follows. A part of the components for ACO are specified in [52]. The artificial ants build solutions: are performed stochastic walks on the completely connected construction. The artificial pheromone trail represents the memory about the ant search process. The pheromone trail is updated during the search process. The quality of a partial solution is given by the amount of pheromone modified along the way. The global solution is found after all ant, based on the previous pheromone information are guided to more promising regions of the search space.

The current approach is using the *Sensitive Ant Model (SAM)* [10] introduced by Chira et al. and further used to solve many complex problems [9–12, 40–44].

The *SAM* model use ants/agents able to communicate indirectly in a stigmergic manner, based on the pheromone trails. In *SAM* the ants/agents are endowed with degrees of heterogeneity in order to improve its search capabilities. The agents are endowed with different levels of sensitivity (pheromone sensitivity level *PSL*) to artificial pheromone and induce different types of reactions to a dynamic environment. The balance between exploring and exploiting could be achieved using both indirect communication and heterogeneous agents.

The *PSL* value is expressed by a real number in the unit interval [0, 1]; when the value is null the agent completely ignores the information and for value 1 it has maximum pheromone sensitivity. When the *PSL* value is low shows that the agents can choose very high pheromone-marked moves, are more independent and are environment explorers; when *PSL* is high the agents are sensitive to pheromone traces and intensively exploit the promising search regions. The learning process produce a modification in time of the pheromone sensitivity: *PSL* increases or decreases based on the search space topology encoded in the ant/agent experience.

The following notations are used: $p_{iu}(t, k)$ is the probability for agent k of choosing the next node u from current node i ; $sp_{iu}(t, k) = p_{iu}(t, k) \cdot PSL(t, k)$ is the re-normalized transition probability for agent k (influenced by *PSL*) where *PSL(t, k)* represents the *PSL* value of agent k at time t. For each node i we have $\sum_u sp_{iu}(t, k) < 1$. The transition probability associated to the virtual state *vs* is $sp_{i,vs}(t, k) = 1 - \sum_u sp_{iu}(t, k)$.

For an agent k at moment t: $sp_{i,vs}(t, k) = 1 - PSL(t, k) \sum_u p_{iu}(t, k)$ and $sp_{i,vs}$ $(t, k) = 1 - PSL(t, k)$. The re-normalized probability $sp_{i,vs}(t, k)$ can be correlated to the system heterogeneity at time t. An interpretation of $sp_{i,vs}(t, k)$ is: the granular heterogeneity of the agent k at iteration t. So, the measure of the system heterogeneity is:

$$E = \sum_k \sum_i (sp_{i,vs}(t, k))^2. \tag{17}$$

The minimum heterogeneity is associated with the maximum sensitivity to pheromone of *SAM* agents and maximum heterogeneity corresponds to *SAM* agents with null sensitivity. The variable E measures the heterogeneity of a *SAM* system versus its corresponding *ACS* variant.

In practice a next move is selected based on the pseudo-random proportional rule *ACS* where the re-normalized transition probabilities *SAM* is considered, mechanism called virtual state decision rule.

In *Medical Image Processing* it is essential for patients life to use very precise mechanisms of image segmentation, edge detection and avoiding noise. The newly introduced strategies in the *Hybrid Medical Image Processing-Sensitive Ant Model* (*HMIP-SAM*) are further described.

4.2 New Strategies Involved in Image Segmentation of Medical Images

In order to improve Image segmentation of a medical image is introduced the heterogeneity factor. In the search strategy the ants use different levels of sensitivity. The *HMIP-SAM* uses different strategies to guide the actions of the ants distributed among the background and target of the image. The image noise needs some special processing.

At first about the background and the target image. The process of segmentation, the clustering process is based in general on [36] and use also heterogeneity.

- The initial cluster centers are the gray values: the number of n peak points with n meanwhile being the gray feature; it reduces the algorithm running time.
- It is set the gradient value of the initial centers to zero. Other gradient values of cluster centers are $gf = \frac{1}{m}(max_{j=1,...n}(\sum_{i=1}^{m} gd(i, j)))$ where $gd(i, j)$ is the gradient value of pixel (i, j), and the image size is $m \times n$.
- It is set the neighborhood eigenvalue for each initial center $ev = 8$. While algorithm is running the neighborhood characteristics of the cluster center value will be 6, and in particular 3 when noise is present (as in [36]).

Now that the initial centers of clusters are found, the algorithm starts to find the best clustering/segmentation of the image. The heuristic function indicating the degree of which the current search pixel is expected to assign to a class, is the similarity. The similarity η of a current search pixels and cluster centers is $\eta_{i,j} = \frac{r}{d(i,j)}$ where r is the cluster radius. Some standard *Ant Colony System* features are further used: at first the same probability to determine the probability that in the target and background areas, ant i at time t transfers to the next pixel j; secondly the globally update pheromone trail. The different levels of sensitivity newly introduced have the purpose to make a difference at the update pheromone phase.

4.3   New Strategies Involved in Edge Detection of Medical Images

In the edge (areas with strong intensity contrasts) detection phase of the medical image processing, the ants moves are based on the differences of the ev-neighborhood pixel value. To improve the search performance of the ant-based model is used the largest adjacent difference value and the maximum connection similarity [15]. The largest adjacent difference value to the edge is the value identified by the ants from the high-neighborhood differences. The maximum connection similarity is when ants find the pixels near the real edge. The heuristic function on the edge is modified as in [36]:

$$\eta_{i,j} = \frac{\sum_{l \in NE_j} |p_j - p_l|}{ev \cdot max\{1, |p_j - p_i|\}}, \tag{18}$$

where p_i is the intensity of pixel i; p_j is the intensity of pixel j next to pixel i; $max\{1, |p_j - p_i|\}$ is the maximum connections similarity factor guiding the ants in searching; NE_j is a set of ev (in particular eight) pixels in the field. In particular, when $\eta_{i,j}$ from Eq. (18) is zero, the ants stop.

Edge detection of an image is a very sensible task. To find a properly one, to not get stuck in the same partial solution based on the indirect communication, the ants are newly endowed with *direct* communication as the agents in Multi-Agent Systems.

The *direct* communication is using a *queen* ant as redirecting information. Each ant could *direct* communicate with the *queen* and the *queen* with each of the ants. The *queen* keeps all the information gathered from the ants and has the possibility to *inform* the ants when necessary. It could be, for example, a potential false edge detection considered to be good, based on its list of noise points obtained by the ants with high PSL or for example to guide ants, especially the ants with low sensitivity level, to further explore the search space.

The global updating rule use as in [36] the average of the mth ant's average step length, and the maximum pheromone in the image. The complex algorithm is executed for a limited time based on a given number of iterations.

When ants detect noise, the noise points are deleted [31]. In the current ant-algorithm *HMIP-SAM* the ants with higher sensitivity are employed to find the noise points. When are found the information is *direct* communicated to the *queen* that will further inform the ants when necessary in the searching process. In this way the accuracy of edge detection is improved.

The scheme of the proposed (*HMIP-SAM*) algorithm, based on [36] follows:

- initialize the ant algorithm parameters and place randomly the ants in the search space;
- initialize each ant with a random sensitivity level;
- start the iteration phase;
- identify the region that the ants are located in through the number of ant ks pixels that have similar pixel gray in a 3 neighborhood.

(1) If an ant k is in the background and target region, it is transferred to this area first. Based on the *ACS* probability of selecting the next element and moves forward. The ants level of sensitivity is modified based on cooperative learning. It is globally update the pheromone trail.

(2) If an ant k with a small PSL is on the edge, calculates the probability of selecting the next element and moves forward according to the state transition probability. The direct communication with the *queen* and the ants with a low PSL stops the over-bounding of pheromone and produce diversification in exploration. The pheromone is updated locally and globally. If the ant has a high PSL it will not identify the edge.

(3) If an ant k with high PSL identify the noise point, the state transition probability is computed and is updated the pheromone on the path. The ants with high PSL direct communicate with the "queen" about noise points; if the ant has a small PSL it will not identify the noise point.

- Stop the iterations when the given number of total iterations is reached.

The result is the best possible solution for a given medical image processing problem, after a given number of trials.

5 Discussion and Further Work

Several discussion about segmentation of medical image follows. We want to under-line the influence of heterogeneity in the current ant-based algorithm. The dynamism of the different level of sensitivity of an ant when running the algorithm is based on ants' learning capabilities influenced by the indirect communication through pheromone trails. The sensitivity factor has the potential to improve the solution of the medical image segmentation problem given in [36].

The *HMIP-SAM* model is using a number of iterations based on the sizes of the images: the larger an image the larger is the number of iterations. This is not also a general rule for the number of ants. The number of ants and the other ant algorithms parameters should be tested on each particular medical image in order to obtain better solution.

The ants have based on their level of sensitivity precise missions: the ants with small PSL will identify better the edges of the image and the ones with higher sensitivity will better identify the noise points. The ants with the smallest PSL are used to identify the medical image edge The direct communication with the *queen* and these ants enhance exploration. Another potential major improvement is the using of the highly sensitive ants to identify the noise points in the edge of the medical image. Using *direct* communication the *queen* is informed about the marked noise point. Further, the *queen* directly inform the other ants in the edge detection process.

In the very recent experiments of Liu et al. it is shown that their approach [36] is better than the one of classical ant-based algorithm that in comparison has "low continuity with severe false and missing detection on edge". In the future will be implemented the proposed *Hybrid Medical Image Processing-Sensitive Ant Model* (*HMIP-SAM*). As a plus, the proposed algorithm *HMIP-SAM* includes heterogeneity and direct communication that have a huge potential to improve the image processing problems and especially medical image (eg. tomography) due to their small details to be discovered. As a future work, novel hybrid models on medical segmentation could be also implemented and tested involving the ant-based clustering model based on similarities as in [45].

6 Conclusion

The paper illustrates some of the efficient image processing tools including ant-based algorithms and in particular models specialized on medical image processing. It is also introduced a new theoretical model based on ants cooperative behavior involving also a *sensitivity* feature and a particular indirect communication inspired by multi-agent systems' communication. Future works will include practical image processing using the *HMIP-SAM* algorithm.

The proposed model *Hybrid Medical Image Processing-Sensitive Ant Model* (*HMIP-SAM)* ensures a balance in image processing. The ants with different levels of sensitivity are used in the segmentation of the image, the ants with small pheromone sensitivity are mainly used in the edge detection phase and the ants with high pheromone sensitivity are mainly used in detection of noise points. The *HMIP-SAM* has a huge potential and could be further improved using for example other multi-agent or robotics characteristics.

References

1. Asbury, C.: Brain imaging technologies and their applications in neuroscience. The Dana Foundation (2011)
2. Asha, A.A., Victor, S.P., Lourdusamy, A.: Feature extraction in medical image using ant colony optimization: a study. Int. J. Comput. Sci. Eng. **3**(2), 714– 721 (2011)
3. Beckmann, E.C.: Br. J. Radiol. **79**, 5–8 (2006)
4. Byrne, C.: Iterative algorithms in tomography. UMass Library (2005)
5. Byrne, C: The EMML and SMART Algorithms. UMass Library (2006)
6. Byrne, C.: Iterative algorithms in inverse problems. UMass Library (2006)
7. Byrne, C.: Applied iterative methods. AK Peters, Wellesley (2008)
8. Cerello, P., et al.: 3D object segmentation using ant colonies. Pattern Recogn. **43**(4), 1476–1490 (2010)
9. Chira, C., Pintea, C.-M., Dumitrescu, D.: A step-back sensitive ant model for solving complex problems. In: Stud Univ Babes-Bolyai Inform KEPT2009, pp. 103–106 (2009)
10. Chira, C., Pintea, C.-M., Dumitrescu, D.: Sensitive ant systems in combinatorial optimization. In: Stud Univ Babes-Bolyai Inform KEPT2007, pp. 185–192 (2007)
11. Chira, C., Pintea, C.-M., Dumitrescu, D.: Sensitive stigmergic agent systems: a hybrid approach to combinatorial optimization. Adv. Soft Comput. **44**, 33–39 (2008)
12. Chira, C., Pintea, C.-M., Dumitrescu, D.: Cooperative learning sensitive agent system for combinatorial optimization. Stud. Comput. Intell. **129**, 347–355 (2008)
13. Crisan, G.-C., Nechita, E.: Solving fuzzy TSP with ant algorithms. Int. J. Comput. Commun. Control Suppl. III, 228–231 (2008)
14. Crisan, G.C.: Ant algorithms in artificial intelligence. Ph.D. Thesis, Al. I. Cuza University of Iasi, Romania (2007)
15. De -Sian, L., Chien, C.C.: Edge detection improvement by ant colony optimization. Pattern Recogn. Lett. **29**, 416–425 (2011)
16. Dorigo, M., Stützle, T.: Ant Colony Optimization. MIT Press, Cambridge (2004)
17. Edholm, P.R., Herman, G.T.: Linograms in image reconstruction from projections. IEEE Trans. Med. Imaging **6**(4), 301–307 (1987)
18. Escalante, R., Marcos R.: Alternating projection methods. SIAM, 8 (2011)
19. Fernandes, C.M., Ramos, V., Rosa, A.C.: Self-regulated artificial ant colonies on digital image habitats. ILCJ **1**(2), 1–8 (2005)
20. Gordon, R., Bender, R., Herman, G.T.: Algebraic reconstruction techniques (ART) for three-dimensional electron microscopy and x-ray photography. J. Theoret. Biol. **29**, 471–481 (1970)
21. Gupta, K.: Image enhancement using ant colony optimization. IOSR J. VSLI Signal Proc. **1**(3), 38–45 (2012)
22. Herman, G.T.: Fundamentals of computerized tomography: Image reconstruction from projection, 2nd edn. Springer (2009)
23. Hornich, H.: A tribute to Johann radon. IEEE Trans. Med. Imaging **5**(4), 169–169 (1968)
24. http://archaeology.tau.ac.il/azekah/
25. http://surfacesearch.com/page11/page3/page4/page4.html
26. http://www.bgs.ac.uk/research/tomography/
27. http://www.britannica.com/topic/tomography

28. http://www.uniongeneralhospital.com/
29. https://en.wikipedia.org/wiki/Ocean_acoustic_tomography
30. https://en.wikipedia.org/wiki/Quantum_tomography
31. Jinghu, Z.: Study on the image edge detection based on ant colony algorithm. Shangxi University (2008)
32. Kaczmarz, S.: Angenäherte auflösung von systemen linearer gleichungen. Bull. Acad. Pol. Sci. **35**, 355–357 (1937)
33. Kaczmarz, S.: Approximate solution of systems of linear equations. Int. J. Control **57**(6), 1269–1271 (1993)
34. Katteda, S.R., Raju, C.N., Bai, M.L.: Feature extraction for image classification and analysis with ant colony optimization using fuzzy logic approach. SIPIJ **2**(4), 137–143 (2011)
35. Liang, Y., Yin., Y.: A new multilevel thresholding approach based on the ant colony system and the EM algorithm. Int. J. Innov. Comput. I **9**(1), 319–337 (2013)
36. Liu, X., et al.: Image segmentation algorithm based on improved ant colony algorithm. Int. J. Signal Proc. Image Proc. Pattern Recogn. **7**(3), 433–442 (2014)
37. Marco, S., Boudier, T., Messaoudi, C., Rigaud, J.-L.: Electron tomography of biological samples. Biochemistry (Moscow) **69**(11), 1219–1225 (2004)
38. Möbus, G., Inkson, B.J.: Nanoscale tomography in materials science. doi:10.1016/S1369-7021(07)70304-8
39. Narayanan, M., Byrne, C., King, M.: An interior point iterative maximum-likelihood reconstruction algorithm incorporating upper and lower bounds with application to SPECT transmission imaging. IEEE TMI **20**(4), 342–353 (2001)
40. Pintea, C-M., Pop, C.P.: Sensor networks security based on sensitive robots agents. A conceptual model. Adv. Intell. Syst. Comput. **189**, 47–56 (2013)
41. Pintea, C.-M.: Advances in bio-inspired computing for combinatorial optimization problem. Springer (2014)
42. Pintea, C.-M., Chira, C., Dumitrescu, D., Pop, P.C.: A sensitive metaheuristic for solving a large optimization problem. LNCS **4910**, 551–559 (2008)
43. Pintea, C.-M., Chira, C., Dumitrescu, D.: Sensitive ants: inducing diversity in the colony. Stud. Comput. Intell. **236**, 15–24 (2009)
44. Pintea, C.-M., Pop, C.P.: Sensitive ants for denial jamming attack on wireless sensor network. Adv. Intell. Soft Comput. **239**, 409–418 (2014)
45. Pintea, C.-M., Sabau, V.: Correlations involved in a bio-inspired classification technique. Stud. Comput. Intell. **387**, 239–246 (2011)
46. Popa, C.: Projection Algorithms-Classical Results and Developments: Applications to Image Reconstruction. LAP, Lambert Academic Publishing (2012)
47. Radon, J.: Über die Bestimmung von Funktionen durch ihre Integralwerte Langs Gewisser Mannigfaltigkeiten [On the determination of functions from their integrals along certain manifolds]. Ber. Verh. Sachs. Akad. Wiss. **69**, 262–277 (1917)
48. Radon, J.: On the determination of functions from their integral values along certain manifolds. IEEE Trans. Med. Imaging **5**(4), 170–176 (1986)
49. Rockmore, A., Macovski, A.: A maximum likelihood approach to emission image reconstruction from projections. IEEE Trans. Nucl. Sci. **23**, 1428–1432 (1976)
50. Salewski, M., et al.: Doppler tomography in fusion plasmas and astrophysics. Plasma Phys. Controlled Fusion **57**, 014021
51. Vardi, Y., Shepp, L.A., Kaufman, L.: A statistical model for positron emission tomography. J. Am. Stat. Assoc. **80**(389), 8–20 (1985)
52. Vescan, A.: Construction approaches for component-based systems. PhD. Thesis. Babes-Bolyai University (2008)
53. Wernick, M.N., Aarsvold, J.N.: Emission tomography: the fundamentals of PET and SPECT. Academic Press (2004)
54. Wu, G., et al.: Geometric correction method for 3d in-line X-ray phase contrast image reconstruction. Biomed. Eng. Online **13**(105) (2014)

Smoke Detection in Environmental Regions by Means of Computer Vision

Douglas A.G. Vieira, Adriano L. Santos, Hani C. Yehia, Adriano C. Lisboa and Carlos A.M. Nascimento

Abstract This paper presents a novel method for smoke detection in videos based on three main steps: background removal, color classification, and temporal/spatial persistence. First, the regions of interest are determined by detecting objects in movement through background subtraction. After that, the moving objects are classified according to their color. Only objects with smoke colors are analyzed by the last stage, which carries out analysis of space movement, verifying the temporal persistence of pixels of interest in the video frames. Pixels that satisfy the rules defined in the three stages are considered pixels of smoke. The system considered in this work is compared with other existing systems, through tests with a database of videos that contains smoke images and videos that do not contain smoke images, but contain objects that look like smoke.

Keywords Computer vision · Smoke detection · Feature selection · Temporal/ spatial persistence

1 Introduction

Fire is one of the main causes of deforestation, destroying a large percentage of woods and forests. Moreover, they are responsible for producing and emitting a great amount of CO_2 into the atmosphere and, on the top of forest destruction, it also

D.A.G. Vieira (✉) · A.C. Lisboa
ENACOM Handcrafted Technologies, Belo Horizonte, Brazil
e-mail: douglas.vieira@enacom.com.br
URL: http://www.enacom.com.br

A.C. Lisboa
e-mail: adriano.lisboa@enacom.com.br

A.L. Santos · H.C. Yehia
Graduate Program in Electrical Engineering,
Federal University of Minas Gerais, Belo Horizonte, Brazil

C.A.M. Nascimento
CEMIG-Companhia Energetica do Estado de Minas Gerais, Belo Horizonte, Brazil

© Springer International Publishing Switzerland 2016
I. Hatzilygeroudis et al. (eds.), *Combinations of Intelligent Methods and Applications*, Smart Innovation, Systems and Technologies 46,
DOI 10.1007/978-3-319-26860-6_8

135

causes ambient unbalance. An efficient solution to reduce and to prevent the damages caused by forest fires is detecting and extinguishing fires as fast as possible, thus, avoiding them to evolve quickly to an uncontrolled fire of great proportions. One manner to reach this objective is creating systems that anticipate fire detection and send alerts to the authorities. Therefore, it is interesting detecting the smoke in the initial fire moment.

Early fire detection allows rapid response and proper fighting against its spreading. This environmental monitoring can be done through smoke detection and alarm generation. One way used to prevent forest destruction caused by fires is monitoring it using videos.

Exploring the characteristics of a charge-coupled device (CCD) camera, an automatic system for fire detection in videos was proposed in [1]. In [2] the image energy was evaluated by means of the Wavelet Transform coefficients and Color Information for smoke detection. An image energy statistical model of it is built using a temporal Gaussian Mixture.

An automatic system for early smoke source detection through the real time processing of landscape images was described in [3]. It explores segmentation to extract persistent dynamical pixels envelopes into the images, together with temporal filtering and spatial analysis. The smoke is finally discriminated from other phenomena using transitory and complex motion.

In [4] it is proposed a method that, after using background removal, computes the frame high frequency energy to define either if the region contains smoke or not. It is based on the assumption that the smoke smooths the images, making it harder to detect edges. This method is further improved in [5]. A real-time fire-detector that combines foreground object information with fire color pixel statistics was proposed in [6]. It considers an adaptive background subtraction algorithm and a color statistical validation to determine the pixels which are fire candidates.

Running in a YCbCr color space, a rule-based generic color model for flame pixel classification is proposed in [7]. It argues that the YCbCr color space separates the luminance from the chrominance more effectively than color spaces such as RGB or rgb. This method was extended in [8] using the CIELab color space to identify fire pixels. Classification rules in the RGB space is presented in [9], while in [7], the YCbCr color space is considered. The YUV color space is used in [10]. The use of feature selection methods in several colors codification is presented in [11].

In [12] the Mortons integral equation was introduced for calculating the maximum plume rise, and beam smoke detectors were recommended for smoke detection design. In [13] it was considered that the optical flow is a good approximation of the motion field. Using the Lucas Kanade's optical flow algorithm, candidate regions are calculated, and it is used to differentiate smoke for other moving objects. The use of video sequences captured by Internet Protocol (IP) cameras was explored in [14]. It has considered features such as color, motion and growth properties using the Discrete Cosine Transform domain to reduce computational cost.

In [15] a four steps method is proposed for smoke detection considering multi-scale analysis, local binary patterns, local binary patterns variance used as features for a neural network classification. In [16] a double mapping framework using AdaBoost is proposed. It considers histogram of edge orientation, edge magnitude, edge densities to help detecting smoke with arbitrary shapes.

This work aims at developing a smoke detection system by means of video sequences, where the purpose of the system is the smoke detection in early stages of formation.

2   The Proposed Method

Recognizing smoke in a video can be a hard task since it has many similarities with clouds and fog, among others, as shown in Fig. 1. Therefore, several features must be taken into account for proper detection. Moreover, the fire fuel changes the smoke behavior, as shown in Fig. 2. This work will explore the white smoke originated from forests in the fire early stage. It is a white smoke due to the high content of water, and it is of interest since it is related to the very beginning of the fire.

The proposed method has the following steps:

1. Background removal;
2. Segmentation of moving regions;
3. Color classification;
4. Region grouping;
5. Temporal/spatial persistence.

The main steps are detailed next.

Fig. 1 Images visually similar to smoke

Fig. 2 Smokes from different types of fuel

2.1   Background Removal

Following the results presented by [17], the chosen algorithm for the background
removal is based on the adaptive mean. In this algorithm, the rate α determines the
influence of the current frame in the background model. The background model B_i
is initialized as the first frame and the next steps are updated as:

$$B_i(x, y) = \begin{cases} f_i(x, y), & \text{if } i = 2 \\ (1 - \alpha)B_{i-1}(x, y) + \alpha f_i(x, y), & \text{if } i > 2 \end{cases}$$

where (x, y) are the pixel coordinates, f_i is the frame i. Afterwards, the moving
objects are defined as the ones such that

$$A_i(x, y) = \begin{cases} 1, & \text{if } |f_i(x, y) - B_i(x, y)| > \tau \\ 0, & \text{otherwise} \end{cases}$$

where τ is a threshold of movement. In Fig. 3 it is presented a video processing
varying α, while in Fig. 4 it is presented the effects of varying τ. The values of these
parameters must be set considering the frame rate and video resolution.

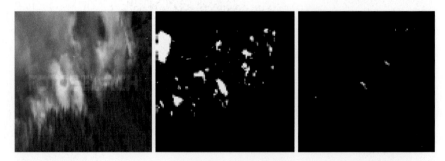

Fig. 3 The original video, and the background removal using $\alpha = 0.0075$ and $\alpha = 0.1$ for $\tau = 50$,
respectively

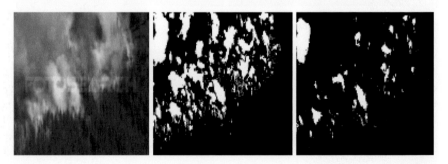

Fig. 4 The original video, and the background removal using $\tau = 25$ and $\tau = 50$ for $\alpha = 0.0075$,
respectively

The α and τ parameters was empirically determined since there is not a standard value for them, as discussed in [18]. Boult et al. discuss video-based target detection, specially the background adaptation techniques [19]. The Lehigh Omni-directional Tracking System (LOTS), adaptive multibackground modeling, quasi-connected components, background subtraction analyses, are analyzed.

2.2 Color Classification

The next step is classifying the moving objects according to their color. The aim of this work is to detect smoke in its early stage, which contains large amount of water. It is characterized as a white/gray smoke. A dataset was built manually to label the smoke regions in some training images, as presented in Fig. 5.

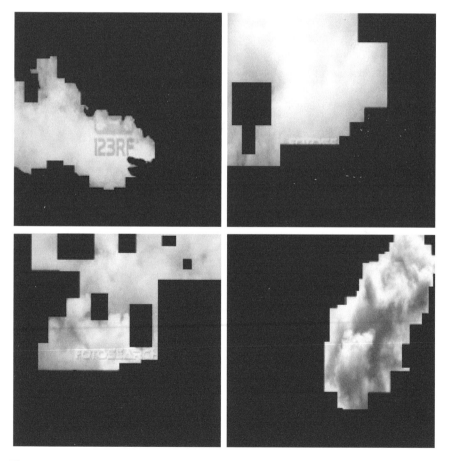

Fig. 5 Sample of the segmented images used in the training set

The smoke pixels were defined as class 1, also called positive or with smoke, and the ones without smoke were defined as class -1, also called negative or smokeless. Observe that a color can be assigned to the presence of smoke in a pixel, and the absence in another pixel.

Only components of color models are considered to develop a classification method by pixel. Its results must be combined to other methods, like motion detection and direction, in order to get an accurate smoke detection. The database used in this paper consists of smoke predominantly white in daylight photos of landscapes, which is incorporated in the knowledge for classification. Other specific databases can be used for specific environmental conditions.

Some colors can be classified as smoke in a pixel and smokeless in another pixel because of the lighting conditions and smoke features. Thus, some colors may receive more than one label, i.e. they may belong to more than one class when they lie on class boundaries. Therefore, it is necessary to use a criterion to remove multiple labeling. Samples with the same label but with lower frequency are then removed in order to preserve the representativeness of the information contained in the original pixels.

The pixels with smoke color were clustered to define the classification using the k-means method. Spherical clusters were considered, as presented in Fig. 6. It was considered 30 clusters to define the positive region.

Figures 7 and 8 present the color classification applied to a frame, where the pixels marked in red are the ones with the color of interest. Note that the dark color smoke was not detected by the classifier.

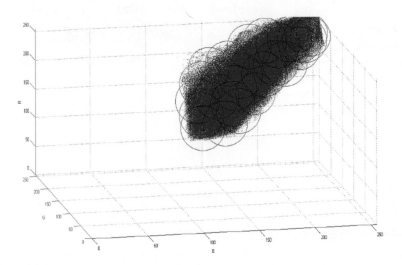

Fig. 6 The classifier in the RGB space. The points in *blue* are the segmented smoke

Fig. 7 An example of smoke color classification in a clear day-light. The pixels in *red* are smoke candidates

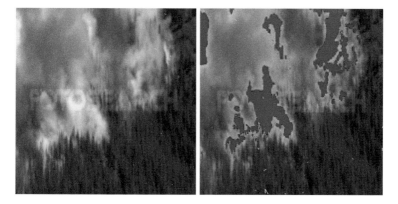

Fig. 8 An example of smoke color classification for a darker smoke in a darker environment. The pixels in *red* are smoke candidates

2.3   Spatial/Temporal Persistence

This step aims at exploring the spatial and temporal characteristics for a set of frames. For doing so, it was elaborated a tracking model applied to the moving objects classified with the smoke color. The spatial temporal characteristic of each cluster is analyzed in a set of frames. The clusters are built considering the pixels connectivity using the find-union strategy.

A spatial clustering is presented in Fig. 9. Each cluster is identified with a given color based on the distance between the classified pixels. Some clusters are very small and filtered from the final result, and they are marked inside circles.

The clusters are followed during a set of frames and, after some persistence, they are labelled as smoke. Figure 10 presents a result considering the persistence, where the ones within the red square are the ones labelled as smoke, and the ones within

Fig. 9 *Each color* in the processed image represents a cluster of smoke pixels. These clusters will be followed during a time to confirm it is smoke. The small clusters marked inside *circles* are ignored due to their small size

Fig. 10 Temporal persistence, where the pixels within the *red square* are the ones labelled as smoke, and the ones within the *yellow square* are candidates that have not yet fulfilled the temporal conditions

the yellow square are candidates that have not yet fulfilled the temporal conditions. The temporal validation is useful to avoid false positive. It was set as 30 frames, therefore, 3 s in the tested videos.

3   Simulated Results

In this section some results are explored. The method proposed in this work was develop in C programming language with the support of the OpenCV (Open Source Computer Vision Library) library [20]. The OpenCV is an open source computer vision and machine learning software with more than 2,500 optimized algorithms.

This section presents some results for the videos applied for testing. Smoke in several conditions is presented. In Fig. 11a it is presented a scene with large amount of dark smoke. A similar situation is presented in Fig. 11b, with the camera focus closer to the region of interest. Since the fire is advancing through the forest, it keeps the white smoke as it gets into green vegetation.

(a)   **(b)**

Fig. 11 Some processed video. The method was set to detect the smoke with water content, the part in *white color*. **a** A video with smoke with large amount of dark content, **b** a video with a closer zoom in the smoke

(a)   **(b)**

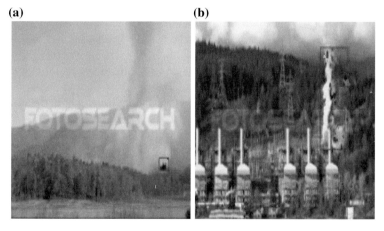

Fig. 12 Some processed video with some different features. **a** Smoke with small content of water, **b** smoke from an industry

Figure 12a considers a more concentrated fire in a evolved movement, thus, with a large amount of dark smoke. The result in Fig. 12b presents a smoke from a factory with large amount of water. It is interesting to note that the clouds were not classified as smoke. It has happened due to the background removal step. This set of videos presented some of the most common aspects in smoke detection in environmental conditions.

3.1 Comparative Results

The obtained results are compared with three well-known methods. The first one is the [9] algorithm. This algorithm was adapted to use background removal, defining our second test method. The third method was the one developed in [4]. Every test presented in this section was made using a 2.97 GHz, 8 GB RAM notebook.

The first data set consists of 10 videos with the following settings: 10 frames/s and 10 s of duration, composing an amount of 100 frames. The data set contains 5 videos without any smoke and other 5 with smoke in every frame.

The videos have the following characteristics, as shown in Fig. 13:

- daylight with white clouds (two videos);
- fog in a forest;
- frozen trees moving in a gray background;
- white boats moving in a river;
- smoke in a forest (four different videos); and
- smoke getting out of a chimney.

Varying the videos resolution, four methods are compared: the proposed method; Algorithm 1 as in [9]; Algorithm 2 it is [9] with background removal; and Algorithm 3 as in [4]. The results are presented in Fig. 14. The resolution was varied from 160×120 up to 1920×1080 (Figs. 15 and 16).

The results show that the proposed method has an interesting performance when compared to the other existing techniques. Moreover, it is clear the importance of the movement detection to avoid that objects with similar colors of smoke to be mistakenly detected, as happened with Algorithm 1 [9]. The results also suggest that higher resolutions can deteriorate the final results. However, depending on the distance from the objects of interest different settings should be used and higher resolution would be appropriate. The rate of false negative classification is presented in Fig. 18.

Fig. 13 Smoke form an industry

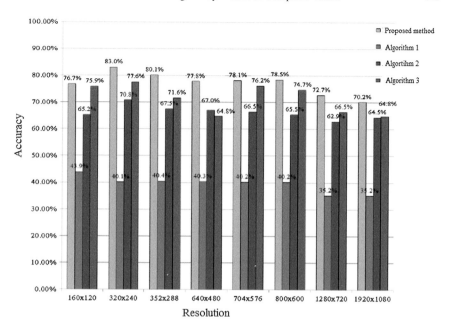

Fig. 14 Accuracy comparison results varying the videos resolution: the proposed method; Algorithm 1 as in [9]; Algorithm 2 it is [9] with background removal; and Algorithm 3 as in [4]

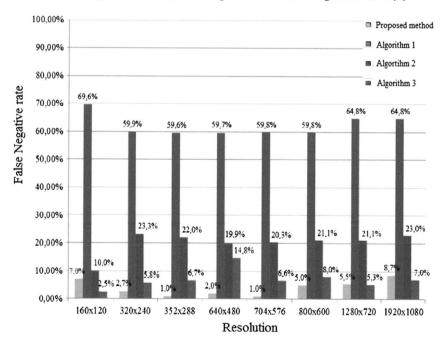

Fig. 15 False negative rate comparison results varying the videos resolution: the proposed method; Algorithm 1 as in [9]; Algorithm 2 is [9] with background removal; and Algorithm 3 as in [4]

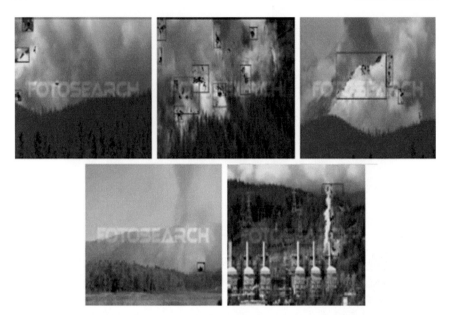

Fig. 16 Experiment 1 results from the videos with smoke processed with the proposed method

A second experiment was held using the data base employed in [4] with the following characteristics, as presented in Fig. 17:

- small smoke spot in a forest (two videos);
- cars headlight during night (two videos);
- distant smoke in a forest;
- smoke in a garden (four videos);
- smoke in a indoors environment (two videos);
- smoke getting out of a chimney;
- smoke in a parking lot.

It is interesting to note the high diversity of images in the videos. In particular the car headlights may cause difficulties for methods that look for fire. The two indoors videos have a white wall as the background, which is likely to be accepted by the color classification. Moreover, the scenes with smoke have good variety of background and smoke distance, and, therefore, it is a good test for the methods (Fig. 18).

The proposed method has an accuracy of 77.18 % while the method from [4] achieved 74.39 %. The proposed method presented 1.30 % of false negatives, while [4] achieved 2.40 %.

Fig. 17 Frames from the second data set extracted from [4]

Fig. 18 Experiment 2 results from the videos with smoke processed with the proposed method

4 The Infrastructure Setting

This project was designed to cover environmental areas in remote regions aiming at generating alarms to support faster protection actions. For doing that, a special infrastructure was design for a real-world testing. The basic equipments used are shown in Fig. 19 and they are composed by a mainboard+ mini-PCI wireless, a camera, a media converter, coolers and a CPU (optional). It must be protect and requires at least 35 W for running.

The power consumption is a complex bottleneck since the aim of this project is to cover large remote regions such as parks, forests and conservation areas.

The first pilot was installed in the Belo Horizonte Technology Park since it is of easy access and it has several conservation areas surrounding it. In this region, the months between May and October compose the dry season. According to the Minas Gerais state government, over 6,000 hectares of conservation areas were destroyed in 2014 by means of fire. It was registered almost 400 occurrences during that year.

In October, 5th, 2014 a fire has taken place in forest which was monitored by the proposed solution. Even though it is in a urban area, there is a large extend of green area which belongs to the Federal University of Minas Gerais, as shown in Fig. 20. On the image top it is drawn a red rectangle which indicates that smoke has been detected in the scene. The fire was located at about 3 Km from the camera, been a challenging situation.

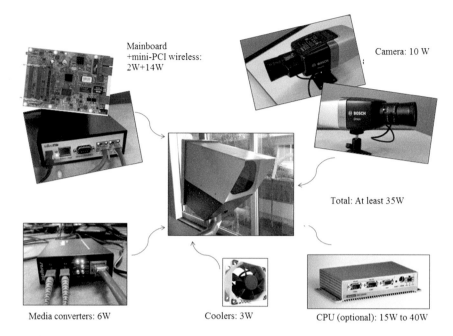

Mainboard
+mini-PCI wireless:
2W+14W

Camera: 10 W

Total: At least 35W

Media converters: 6W   Coolers: 3W   CPU (optional): 15W to 40W

Fig. 19 The set of equipments applied in the real-world experiments

Fig. 20 The set of equipments applied in the real-world experiments

5   Conclusions

This paper has presented a strategy for early detection of white smoke in videos. This strategy is very useful since the first visible signals in a fire for green vegetation is the white smoke. The method has performed well in the tested videos. The system is now in test in a real world situation with two cameras monitoring a small preservation area. As a further improvement, the method should include night light situations. However, during the night it is very simple to detect the fire using its color contents and shine.

Acknowledgments   This work was developed under an R&D project financed by CEMIG Distribution under the number D383, regulated by the Brazil Agency of Electrical Energy (ANEEL - Agencia Nacional de Energia Eletrica).
The authors also would like to thank CNPq, FAPEMIG and CAPES for the financial support.

References

1. Cheng, X., Wu, J., Yuan, X., Zhou, H.: Principles for a video fire detection system. Fire Saf. J. **33**(1), 57–69 (1999). http://www.sciencedirect.com/science/article/pii/S0379711298000472
2. Calderara, S., Piccinini, P., Cucchiara, R.: Vision based smoke detection system using image energy and color information. Mach. Vis. Appl. **22**, 705–719 (2011). doi:10.1007/s00138-010-0272-1. http://dx.doi.org/10.1007/s00138-010-0272-1
3. Vicente, J., Guillemant, P.: An image processing technique for automatically detecting forest fire. Int. J. Thermal Sci. **41**(12), 1113–1120 (2002). http://www.sciencedirect.com/science/article/pii/S1290072902013972
4. Toreyin, B., Dedeoglu, Y., Cetin, A.E.: Wavelet based real-time smoke detection in video. In: Signal Processing: Image Communication, EURASIP (2005). http://www.cs.bilkent.edu.tr/yigithan/publications/eusipco2005.pdf
5. Toreyin, B.U. Dedeoglu, Y., Cetin, A.E.: Contour based smoke detection in video using wavelets. In: 14th European Signal Processing Conference (EUSIPCO 2006) (2006)
6. Celik, T., Demirel, H., Ozkaramanli, H., Uyguroglu, M.: Fire detection using statistical color model in video sequences. J. Vis. Comun. Image Represent **18**(2), 176–185 (2007). http://dx.doi.org/10.1016/j.jvcir.2006.12.003
7. Celik, T., Demirel, H.: Fire detection in video sequences using a generic color model. Fire Saf. J. **44**, 147–158 (2009)
8. Celik, T.: Fast and efficient method for fire detection using image processing. ETRI J. **32**, 881–890 (2010)
9. Chen, T.-H., Yin, Y.-H., Huang, S.-F., Ye, Y.-T.: The smoke detection for early fire-alarming system base on video processing. In: IIH-MSP '06. International Conference on Intelligent Information Hiding and Multimedia Signal Processing, 2006, pp. 427–430 (2006)
10. Marbach, G., Loepfe, M., Brupbacher, T.: An image processing technique for fire detection in video images. Fire Saf. J. **41**, 285–289 (2006)
11. Miranda, G.M.T., Lisboa, A., Vieira, D.A.G., Queiroz, F., Nascimento, C.A.M.: Color feature selection for smoke detection in videos. In: 12th International Conference on Industrial Informatics (2014)
12. Fang, J., Yuan, H.Y.: Experimental measurements, integral modeling and smoke detection of early fire in thermally stratified environments. Fire Saf. J. **42**,(1), 11–24 (2007). http://www.sciencedirect.com/science/article/pii/S0379711206000920

13. Chunyu, Y., Jun, F., Jinjun, W., Yongming, Z.: Video fire smoke detection using motion and color features. Fire Technol. **46**, 651–663 (2010). doi:10.1007/s10694-009-0110-z. http://dx. doi.org/10.1007/s10694-009-0110-z

14. Millan-Garcia, L., Sanchez-Perez, G., Nakano, M., Toscano-Medina, K., Perez-Meana, H., Rojas-Cardenas, L.: An early fire detection algorithm using ip cameras. Sensors **12**, 5670–5686 (2012)

15. Yuan, F.: Video-based smoke detection with histogram sequence of lbp and lbpv pyramids. Fire Saf. J. **46**(3), 132–139 (2011). http://www.sciencedirect.com/science/article/pii/S0379711211000026

16. Yuan, F.: A double mapping framework for extraction of shape-invariant features based on multi-scale partitions with adaboost for video smoke detection. Pattern Recognit. **45**(12), 4326–4336 (2012). http://www.sciencedirect.com/science/article/pii/S0031320312002786

17. Piccardi, M.: Background subtraction techniques: a review. In: 2004 IEEE International Conference on Systems, Man and Cybernetics, vol. 4, pp. 3099–3104 (2004)

18. Collins, R.T., Lipton, A.J., Kanade, T., Fujiyoshi, H., Duggins, D., Tsin, Y., Tolliver, D., Enomoto, N., Hasegawa, O., Burt, P., Wixson, L.: A system for video surveillance and monitoring. The Robotics Institute, Carnegie Mellon University, Technical Report (2000)

19. Boult, T., Micheals, R., Gao, X., Eckmann, M.: Into the woods: visual surveillance of non-cooperative and camouflaged targets in complex outdoor settings. Proc. IEEE **89**, 1382–1402 (2001)

20. (2015). http://opencv.org/

Printed in the United States
By Bookmasters